Visual Phenomenology

Visual Phenomenology

Michael Madary

The MIT Press
Cambridge, Massachusetts
London, England

© 2017 Massachusetts Institute of Technology

All rights reserved. No part of this book may be reproduced in any form by any electronic or mechanical means (including photocopying, recording, or information storage and retrieval) without permission in writing from the publisher.

MIT Press books may be purchased at special quantity discounts for business or sales promotional use. For information, please email special_sales@mitpress.mit.edu or write to Special Sales Department, The MIT Press, 1 Rogers Street, Cambridge, MA 02142.

This book was set in Stone Sans and Stone Serif by Toppan Best-set Premedia Limited. Printed and bound in the United States of America.

Library of Congress Cataloging-in-Publication Data

Names: Madary, Michael, author.
Title: Visual phenomenology / Michael Madary.
Description: Cambridge, MA : MIT Press, [2016] | Includes bibliographical references and index.
Identifiers: LCCN 2016018209 | ISBN 9780262035453 (hardcover : alk. paper)
Subjects: LCSH: Perception (Philosophy) | Visual perception. | Phenomenology.
Classification: LCC B828.45 .M335 2016 | DDC 121/.35–dc23 LC record available at https://lccn.loc.gov/2016018209

10 9 8 7 6 5 4 3 2 1

For Sheila and the girls

Contents

Preface xi
Acknowledgments xiii
Abbreviations xv

Part I 1

1 Introduction 3
1.1 The Main Argument 3
1.2 The Sandwich or the Cycle? 5
1.3 Same Strategy, Different Results 12

2 Three Constraints 27
2.1 Visual Experience Is Perspectival 28
2.2 Visual Experience Is Temporal 32
2.3 Visual Experience Is Indeterminate 36
2.4 Thesis AF and the Three Constraints 38

3 Anticipation and Fulfillment 41
3.1 (PC) and Siegel's Doll 41
3.2 (PC′) and Five Points about Anticipation 43
3.3 Variation in Perceptual Content 54
3.4 Visual Anticipation and Two Distinctions 56
3.5 Summary 57

4 The Question of Content 59
4.1 Introducing AF Content 59
4.2 Alternative Theories of Content and Their Shortcomings 65
4.3 On the Denial of Perceptual Content 70

4.4 Four Problems and Three Solutions   75
4.5 Summary   86

Part II   89

5   Some Perceptual Psychology   91

5.1 Various Strands of Support   92
5.2 Rejecting the Myth of Full Detail   96
5.3 The Importance of Action   99
5.4 Facing the Resistance   103
5.5 Visual Attention   111
5.6 Objections and Replies   117
5.7 Summary   118

6   The Active Brain   119

6.1 Ongoing Cortical Dynamics   119
6.2 Neural Feedback   123
6.3 Theoretical Options   125
6.4 Summary   128

7   The Dorsal Stream and the Visual Horizon   131

7.1 Visual Consciousness and the Two Streams   131
7.2 Introducing the Visual Horizon   134
7.3 Input to the Dorsal Stream   136
7.4 Localized Damage and Illusions   138
7.5 Disturbances of Visual Motion   145
7.6 Computational Models of Dorsal Anticipation   149

Part III   153

8   The Convergence   155

8.1 Back to the Main Argument   155
8.2 The Best of Both Worlds—Symbolic Dynamics   160
8.3 Do We Need Internal Representations?   162

9   Seeing Our World   165

9.1 AF Content Is of a Shared Social World   165
9.2 Empirical Support   170
9.3 Embedded Rationality   174

Appendix: Husserl's Phenomenology   177
A.1 Finding AF in Husserl   177
A.2 Descriptive Psychology or Transcendental Phenomenology?   180
A.3 Phenomenology and the Sciences of the Mind   184

Notes   191
References   203
Index   241

... *die Intentionalität ist vorwiegend der Zukunft zu gerichtet.*
—E. Husserl (Husserliana XI: 156)

... all brains are, in essence, anticipation machines.
—D. Dennett (1991: 177)

Preface

This book covers two different ways of investigating visual experience. The first way is through careful philosophical description of experience. The second way is through the empirical sciences of the mind. The main motivation behind the book is the exciting fact that these two different ways of investigating visual experience have independently converged on the same result. The result is the main thesis of the book: *visual perception is an ongoing process of anticipation and fulfillment.*

The first part of the book will make the case for my thesis on the basis of first-person description (for the most part), and the second part of the book will turn to the empirical evidence. In the third part of the book, I will show how the general framework can be applied in a fruitful manner. Here I should note that the very fact that there is a convergence from historically distinct methodologies makes a first point in favor of the result.

The book is written for philosophers and scientists, and anyone interested in human visual experience. Advanced undergraduate students should find it accessible. I have attempted to create an exchange between the humanities and the sciences in which results from both modes of investigation are taken seriously. At some points I engage with the contemporary philosophical literature to an extent that might test the patience of readers who are not professional philosophers. Such readers are invited to skim or skip the philosophical fine points, as doing so should not take away from a general understanding of the book.

In addition, I should be explicit that there are a number of philosophical issues that currently receive a good bit of attention, but that I will not be addressing, at least not directly. These include metaphysical questions about the relationship of the mind to the body (Armstrong 1999); the mystery of consciousness, including the "hard problem" (Chalmers 1995) and the explanatory gap (Levine 1983); the question of whether visual perception is direct or indirect (Smith 2002); and the question of what determines the content of visual representations (Cummins 1989). Readers should not expect an answer to the question of how the brain gives rise to visual experience, or the question of precisely where, within the brain, visual experience is located (Metzinger 2000; see section 1.3 below). Instead, I make the case that visual experience has a particular intelligible structure, and that empirical evidence suggests a similar structure within the information processing of the visual system. My hope is that limited explanatory ambition will engender surefooted progress.

Acknowledgments

Some of the main themes in this book are ones that I have been working on for several years, and I have had the good fortune of being helped a great deal along the way. I would like to thank Adrian Alsmith, Harald Atmanspacher, Radu Bogdan, Thiemo Breyer, Robert Briscoe, Bruce Brower, Andy Clark, Maxime Doyon, Nivedita Gangopadhyay, Alistair Isaac, Jakob Hohwy, the late Susan Hurley, Thomas Metzinger, Alva Noë, Lisa Quadt, Susanna Schellenberg, Susanna Siegel, Mel Slater, Finn Spicer, Jennifer Windt, Dan Zahavi, the Mainz Journal Club, and two anonymous reviewers for the MIT Press. Very special thanks go to Sascha Fink, Wanja Wiese, and Jeffrey Yoshimi, all of whom offered detailed comments on the penultimate draft of the book. Research for this work was supported by the EC Project VERE, funded under the EU 7th Framework Program, Future and Emerging Technologies (Grant 257695).

Chapter 3 includes material from my "Anticipation and Variation in Visual Content" (2013), *Philosophical Studies 165*, 335–347.

Chapter 7 includes material from my "The Dorsal Stream and the Visual Horizon" (2011), *Phenomenology and the Cognitive Sciences 10*, 423–438.

Chapter 9 includes material from my "Seeing Our World" (2015), in M. Doyon and T. Breyer (Eds.), *Normativity in Perception* (New York: Palgrave Macmillan).

The appendix includes material from my "Husserl on Perceptual Constancy" (2012), *European Journal of Philosophy 20*, 145–165.

All of this material has been reworked, in many places extensively.

Abbreviations

AF: The main thesis of the book; that is, visual perception is an ongoing process of anticipation and fulfillment.

EVM: The early vision module.

F: Factual content. Visual perception represents factual properties, which are properties that are in principle perceivable from multiple perspectives.

PC: Perspectival connectedness. If S substantially changes her perspective on o, her visual phenomenology will change as a result of this change (from Siegel 2010a: 179).

PC': Perspectival connectedness'. If S substantially changes her perspective on o, her visual phenomenology will present different views of o's factual properties.

SA: Specific anticipation. Visual anticipation is more specific than indicated in the consequent of PC.

VCS: Main thesis of chapter 9; that is, visual content has a strong social element (for humans).

Part I

1 Introduction

1.1 The Main Argument

What is the general structure of visual experience? My answer to this question is the main thesis of this book (abbreviated AF):

AF: Visual perception is an ongoing process of anticipation and fulfillment.

But what does AF mean? Is it a claim about the phenomenology, the "what it is like," of visual experience, or is it a claim about the physical processes that enable visual experience? It is both. I mean to claim that the general structure of anticipation and fulfillment describes both the phenomenology as well as the causal processes that enable vision. The Main Argument of the book can be expressed as follows.

The Main Argument
(1) *The descriptive premise*: The phenomenology of vision is best described as an ongoing process of anticipation and fulfillment.
(2) *The empirical premise*: There are strong empirical reasons to model vision using the general form of anticipation and fulfillment.
(AF) *Conclusion*: Visual perception is an ongoing process of anticipation and fulfillment.

Chapters 2 through 4 make the case for the descriptive premise, and chapters 5 through 7 make the case for the empirical premise. Chapter 8 focuses on theoretical issues surrounding the Main Argument itself. The final chapter of the book shows how AF motivates the claim that there is a strong social element to perceptual content[1] for humans. Here I should also note that my defense of AF throughout the book will touch on many of the major issues in the philosophy and sciences of visual perception.

Thus, readers can also approach the book as a survey of those major issues.

The Main Argument is intended to express, in a concise manner, the convergence of results from two distinct ways of investigating visual perception. Both premises make for substantial and, I hope, interesting claims on their own, no matter what one thinks of AF. The reason why I have constructed the argument in this way is to show that the conclusion is supported by two different methods of investigating the mind. Both premises are concerned with human visual experience in the actual world; the descriptive premise is not a modal claim about how vision might be structured in other possible worlds. General features of the Main Argument, along with related issues, will be addressed in more detail in chapter 8, after I make the case for premises (1) and (2).

Perhaps the best way to begin making the case for AF is to place it within the context of more familiar ideas, to situate it within the existing literature. As many readers will know, there are various philosophical and scientific approaches to the study of the mind. In this introductory chapter I hope to place my own strategy within existing approaches, without taking the time to develop a full treatment of these larger issues. The themes raised in this first chapter will be revisited in more detail in the following chapters. In the next section of this chapter, I will discuss the relationship between perception, action, and cognition, and signal ways in which my approach departs from elements of the traditional framework in cognitive psychology. In the third part of the chapter, I will compare and contrast my own argument with a previous attempt by Ray Jackendoff (1989) and, more recently, Jesse Prinz (2012) to find a convergence between visual phenomenology, on one hand, and empirical models on the other. There I will offer preliminary reasons for departing from elements of the historically dominant framework in cognitive neuroscience.

Before entering into relatively recent issues in the philosophy of mind and psychology, I should acknowledge a major intellectual debt behind the descriptive motivation for AF and behind premise (1). I first discovered (1) in Edmund Husserl's work on perception from the early twentieth century, and, as the reader will see, I have appropriated some of Husserl's ideas at various stages in the book. In the final chapter of the book, more recent work in the Husserlian tradition is also taken up as one of the main themes. I have included an appendix for readers who are interested in the

Introduction 5

details about how the content of the book relates to Husserl's philosophical project. Also note that I have usually cited German publications of Husserl's work, but that the section numbers cited (by §) will correspond to the English translations.

1.2 The Sandwich or the Cycle?

In this section, my goal is to situate my work within the contemporary landscape in the philosophy of psychology. The orthodoxy, dating back to the beginnings of cognitive science in the last few decades of the twentieth century, and perhaps back at least to Descartes before that, is to focus on mental states abstracted away from bodily or contextual details. In this approach, human behavior is driven mainly by cognition, which is usually explained in terms of propositional attitudes such as beliefs and desires. Jerry Fodor's language of thought hypothesis (1975, 1987) is a prime example of such an approach, which is commonly called "classical cognitivism." Susan Hurley labeled this approach to the mind the "classical sandwich" (1998: 20). On this view, perception serves as the input to cognition, and action is the output from cognition (see figure 1.1). Cognition is "sandwiched" between perception and action. Thus, perception and action have very little to do with one another, and they play no major role in support of intelligent behavior. Perception and action are merely peripheral, while cognition is central.

One way of departing from the orthodoxy is to focus on the role of perception and action for intelligent behavior. Historically, this focus can be found both in Aristotle (*De Anima*, Shields 2015) and, beginning in the early twentieth century, in the Husserlian phenomenological tradition (Husserl 1973b, Merleau-Ponty 1945/1962). In the last few decades, the focus on the link between perception and action has gained ground both

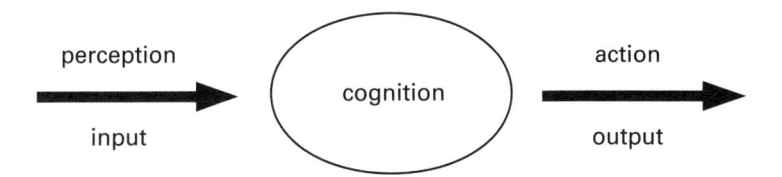

Figure 1.1
The classical sandwich model of the mind (based on Hurley 1998).

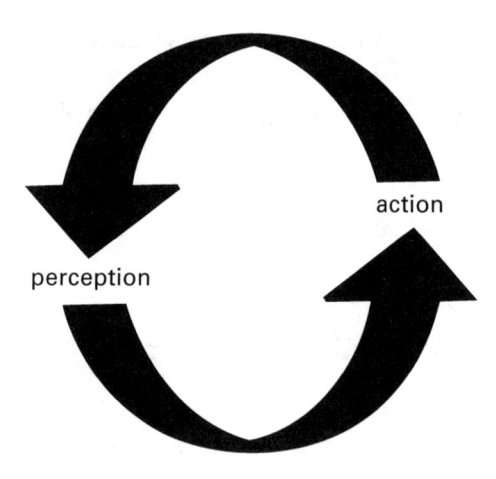

Figure 1.2
The cycle of action and perception.

in philosophical (Hurley 1998, Noë 2004, Schellenberg 2008) and empirical research (Gibson 1979, Prinz and Hommel 2005, Cutsuridis, Hussain, and Taylor 2011). The focus on perception and action marks a departure from the classical sandwich by taking perception and action to be somehow interdependent. Perception and action create a *cycle* of information or, as it is sometimes put, a sensorimotor loop. To put the point roughly—more details will be filled in later in the book—we act by moving and focusing our eyes, which enables perception, which then leads to another action (another eye movement, perhaps) which then immediately creates access to new perceptual information, which then leads to further action, and so on (see figure 1.2).

My own position tends toward the second approach—the action-perception cycle—more than the first. No one can reasonably deny the weak claim that human visual perception is typically an active process of exploring the environment. John Findlay and Iain Gilchrist make the point nicely:

A Martian ethologist observing humans using their visual systems would almost certainly include in their report back: "they move these little globes around a lot and that's how they see." (2003: 1)

In chapter 5, I will present the relevant empirical evidence for their claim in more detail. For now I only note the uncontroversial fact that humans typically perform ballistic eye movements, or saccades, several times per second (Findlay and Gilchrist 2003)—we move our little globes around a lot.

Introduction

While it is not reasonable to deny that human visual perception is typically a process of active exploration, it is reasonable to deny two stronger claims about the relationship between action and perception. The first strong claim is that all human perception, including nonvisual modalities, typically involves active exploration, and the second strong claim is that human visual perception necessarily involves active exploration. The first strong claim faces obvious counterexamples. It is not unusual to enjoy auditory perception, for instance, without actively exploring one's environment. One can be still while, for instance, listening to music. Counterexamples to the second strong claim are not as easily found. Experiments have shown that we are unable to keep our eyes perfectly fixed; even when we try, we still perform small involuntary eye movements (Martinez-Conde 2009). When the motion of the eye is counterbalanced using precise devices to stabilize the retinal image, subjects experience temporary blindness (Riggs and Ratcliff 1952). Thus, the second strong claim, the claim that action is necessary for human visual perception, does have some good empirical support. Alva Noë, who has focused on action and perception in some of his work (O'Regan and Noë 2001, Noë 2004), has been interpreted as being committed to the second strong claim.[2] In response to this reading of Noë, Ken Aizawa has presented possible empirical counterexamples to the second strong claim. One good candidate for a counterexample are cases in which patients report having visual perceptual experiences despite being in a drug-induced total paralysis for the purpose of surgery (Aizawa 2007). Also, in their seminal work on the sensorimotor approach to perception, Kevin O'Regan and Noë discuss experimental conditions involving visual perception without eye movements (2001: section 4.5). They mention experiments in which visual images have been flashed at a rate faster than subjects are able to saccade (under 150 msec), yet subjects are able to perceive the images. O'Regan and Noë regard these cases as exceptions to the natural exploratory nature of visual perception, and they note that, under those conditions, subjects are only able to perceive images that are already familiar to them.

To summarize, we have considered the following three claims about perception and action.

Weak claim: Human visual perception is typically a process of active exploration.

First strong claim: All human perception (including nonvisual modalities) is typically a process of active exploration.

Second strong claim: Human visual perception is necessarily a process of active exploration.

The weak claim is, I take it, undeniable.

The first strong claim is, I take it, false.

The second strong claim is well-motivated, but it is also questionable. I will not take a stance either way.

The thesis of this book, AF, is, in a way, a response to these three claims. The weak claim is too weak, the first strong claim is false, and the second strong claim is quite possibly false as well because of Aizawa's counterexamples mentioned above. The weak claim is too weak for two reasons. First, it offers no hope of generalizing to other perceptual modalities (since the first strong claim is false). It is only a claim about the contingent features of one of our sense faculties and thereby offers no insight for the nature of perception in general. Second, and more importantly, it is only a claim about "typical" visual perception. When we try to universalize it, we run into the objections to the second strong claim. AF includes the weak claim but it also could be generalized to nonvisual modalities, and it is a universal claim about all human visual experience.

When I claim that human visual experience is an ongoing process of anticipation and fulfillment, I mean to include anticipation of the sensory consequences of self-generated movements. Thus, the "typical" case of vision through active exploration—the weak claim—is subsumed under AF. But unlike the weak claim, my thesis could be generalized to other modalities, and it can be universalized to cover all visual experience. Because AF is not committed to a claim about action, only a claim about anticipation, the obvious counterexamples to the first strong claim do not apply. Being perfectly still while listening to a piece of music does not hinder one's ability to anticipate, perhaps implicitly, how a melody will develop. Likewise, because AF is not committed to anything about actual motor movements, the counterexamples to the second strong claim do not apply. Patients who are totally paralyzed could still anticipate the next visual appearance through predictive neural activity alone. By taking the weak claim seriously without being committed to the first and second strong claims, my view includes many of the important insights found in enactivist approaches to

Introduction 9

perception, such as Noë's, without embracing some of more controversial claims that have been associated—fairly or not—with enactivism.

In short, perception is best understood as an ongoing cycle; but instead of a cycle of action and perception, it is better understood using the more general framework of anticipation and fulfillment. In the case of visual perception, the anticipation is "typically" of the consequences of self-generated movements. I will not make the case here that my thesis generalizes to all other sense modalities, but I suspect that it does. The descriptive arguments can be adapted to other modalities,[3] and many of the results from neuroscience that support my thesis are not confined to the visual brain.

Now that I have sketched the way in which AF relates to the action-perception cycle, I should return to the classical sandwich. Above I claimed that it is not reasonable to deny that human visual perception is typically a process of active exploration. Under normal conditions, action and visual perception form an ongoing cycle of information. Does this fact alone threaten the classical sandwich? In order to answer this question, it will be helpful to return to Hurley's discussion of the topic. The substance of Hurley's distinction between the sandwich and the cycle is one of architecture. The sandwich describes a vertical modularity, and the cycle describes a horizontal modularity.

Many readers will be familiar with what Hurley calls vertical modularity from Fodor's influential work (1983). On that view, perceptual systems have nine main features. They are domain-specific, possess mandatory operation, and have limited central access; they are fast, informationally encapsulated, have "shallow" inputs, and are associated with a fixed neural architecture; and they exhibit characteristic breakdown patterns, while their ontogeny exhibits a characteristic pace and sequencing (1983: chapter 3). The general idea is that Fodorian modules produce representations of a particular format, which are then passed on toward a central processor and processed according to rules. The central processor then sends an output in the form of a motor command. Importantly, information is processed in one direction.

Now consider the alternative architecture. Horizontal modularity, according to Hurley,

> sees perception and action as interdependent because codependent on a complex dynamic system of causal relations, which may extend into the environment ... Each horizontal module or layer is a content-specific system that loops dynamically

through internal sensory and motor processes [as] well as through the environment. (1998: 21)

We can distinguish a number of separable, though related, claims within Hurley's notion of horizontal modularity. There is the claim that perception and action are interdependent, the claim that each module is content-specific, and the claim that there are internal and external feedback loops. It would take us off track to investigate each of these claims, and I will return to this theme later on, in the final chapter of the book (section 9.3). For now, though, I will discuss two points of tension between the sandwich and the cycle. The first is the direction of information processing, and the second is the explanation of rationality using propositional attitudes.

The first, and clearest, tension between the sandwich and the cycle is a matter of the direction of information flow. My thesis is incompatible with any view for which information is processed strictly in one direction, from perception, to cognition, to action. As will become clear later, the evidence for massive neural feedback connectivity in cortex is an important part of the support for (2), and thus the empirical support for AF. As some readers will be aware, though, it is not uncommon these days to find models of cognitive architecture that give a central role to some form of feedback (see, for example, Eliasmith and Anderson 2003: chapter 9 or Grush 2004). In this regard, my departure from the classical sandwich is not terribly novel or controversial. But there is a more interesting tension between the cycle and the sandwich.

The more interesting tension has to do with the explanation of rational behavior. Classical cognitivism excels at explaining rational behavior, and it does this explaining, for the most part, in terms of propositional attitudes, at least in philosophical circles (Fodor 1987, Schroeder 2006). It is not clear how an ongoing cycle of perceptual (and typically motor) processing with feedback loops at multiple time scales makes any room for propositional attitudes, or even rationality. Hurley was aware of this problem, and called for a rethinking of the nature of rationality:

Rationality might instead emerge from a complex system of decentralized, higher-order relations of inhibition, facilitation, and coordination among different horizontal layers, each of which is dynamic and environmentally situated. (1998: 409)

With this suggestion, Hurley begins to describe a cognitive architecture that is finding more and more empirical support in recent years. In the final chapter of the book, I will present the various lines of support for this way of understanding rational behavior as it connects visual perception with social cognition. Thus, one theme of the book is that there are good reasons to rethink the traditional view of perception as something distinct from, and having little to do with, cognition.

Of course, increasing support for something like Hurley's idea of horizontal modularity is not alone sufficient to displace propositional attitude attribution as the reigning manner of describing rational behavior within the philosophy of psychology. But it does offer what appears to be an alternative explanation for at least some kinds of intelligent behavior. My main goal in this book is to offer a general framework for understanding human visual perception and, in the last chapter of the book, to show how far this framework can go in explaining some kinds of rational behavior. It would detract from this goal if I were to explore the general issue of perception-action explanation versus propositional attitude explanation. Instead of exploring this question, I will take up a pragmatic strategy partly inspired by Daniel Dennett's approach to this topic (and partly inspired by Anthony Chemero's appropriation of Dennett's strategy [2009: chapter 4]).

Some readers will already be aware of Dennett's various "stances" as ways of predicting the behavior of various kinds of systems (1987: chapter 2). For example, the "physical stance" would involve using information about the detailed physical states of a system, in addition to one's knowledge of the laws of physics, in order to make predictions about the behavior of a system. Dennett, of course, advocates the "intentional stance" as the best strategy for predicting the behavior of rational systems such as us:

Here's how it works: first you decide to treat the object whose behavior is to be predicted as a rational agent; then you figure out what beliefs that agent ought to have, given its place in the world and its purpose.

Then you figure out what desires it ought to have, on the same considerations, and finally you predict that this rational agent will act to further its goals in light of its beliefs. (1987: 17)

This passage expresses one of the main strategies for understanding cognition via propositional attitude attribution. And, quite reasonably, many philosophers and scientists have agreed that this strategy, or something very similar, is effective, although there are important voices of dissent (Churchland 1981, Nichols and Stich 2003: 142–148).

The value of the intentional stance, according to Dennett, lies in its ability to make accurate predictions with limited information. He makes this point by use of a thought experiment in which an Earthling and a Martian enter into a predicting contest. The Martian is a Laplacian super-scientist who is able to predict human activity using the physical stance, using, say, his knowledge of microphysics. The Earthling, on the other hand, uses the intentional stance to make his predictions. Both the Earthling and the Martian are able to predict an instance of human behavior accurately, but the Earthling is able to do so using far less information (1987: 26–27).

What we can take from Dennett is that one could, in principle, use either the physical stance or the intentional stance in order to predict rational behavior. In the final chapter of the book, I will explore another way of predicting and explaining rational behavior, a way that gives a privileged role to visual perception. Although this alternative is not meant to invalidate or rival the intentional stance—just as the intentional stance does not invalidate the physical stance—I suggest that there are cases in which the perceptually based alternative is more natural, elegant, and fruitful than other strategies, than the intentional stance in particular. Such cases offer a basis for understanding how rational behavior is embedded within the cycle of action and perception.

1.3 Same Strategy, Different Results

My general strategy is to explore vision using two different methods. The first method is by careful description of visual experience. The second method is that of natural science. The main point of this book is that the results converge on a similar structure. This strategy is not entirely new. A version of it can be found in work by Ray Jackendoff (1989) and Jesse Prinz (2012).[4] Like mine, their strategy involves making an empirical claim about visual processing as well as a descriptive claim about visual experience. Their empirical claim is that David Marr's theory of visual processing, or something fairly similar to it, is correct. Their descriptive claim is that our conscious visual experience is most like the kind of representation that Marr called the 2.5D sketch. They bring these two claims together by concluding that the 2.5D sketch determines the content of visual experience. Prinz also argues that the location of the 2.5D sketch, and thus the location of visual consciousness, is in the extrastriate cortex in humans. There is a good deal of overlap between my approach and the Jackendoff/Prinz approach. In this section I cover both the overlap and the differences. The main differences are that I think we can and should have a much richer description of perceptual phenomenology and that we ought to accommodate some discoveries from neuroscience that are neglected, or glossed over, in Prinz's account.

In order to reconstruct their general argument, I should first outline the theoretical work which is the basis for their approaches, namely David Marr's remarkably influential theory of vision (1983). His theory is a good example of the classical sandwich mentioned above. Visual perception begins with the retinal image, and then this image is processed in a hierarchical fashion until a full 3D representation of the visual scene is generated and then made available for cognitive processing. His theory has two important methodological presuppositions. First, in line with the classical sandwich, his theory is strictly feedforward. Information flow is in one direction, from the retina onwards up the hierarchy. Second, his method involves abstracting away from what he calls the "implementational level" of explanation. According to Marr, the science of the mind in the second half of the twentieth century revealed that close physiological analysis has serious methodological shortcomings:

The message was plain. There must exist an additional level of understanding at which the character of the information-processing tasks carried out during perception are analyzed and understood in a way that is independent of the particular mechanisms and structures that implement them in our heads. (Marr 1983/2010: 19)

Philosophers will recognize this suggestion as a version of functionalism in the philosophy of mind (Putnam 1975, Lewis 1981, Levin 2013). The idea is that it will be more scientifically fruitful to consider information processing in abstraction from some of the details of the wetware that happens to implement it. (Of course, the challenge is to specify which details can be safely ignored.)

With the basic methodology in place, the core of Marr's theory is that there are three distinct representational stages in visual processing. After the retinal image, the primal sketch is generated, which is devoted to the intensity changes and geometrical organization of the retinal image. Then, the 2.5D sketch is created, which specifies the orientation and depth of visible surfaces. Finally, the 3D model representation is created, which adds full volumetric information to the 2.5D sketch. Much of Marr's groundbreaking book is devoted to developing mathematical techniques for extracting this information from an image in order to create these different levels of representation.

Now here is what I will call the *Jackendoff/Prinz argument* that human visual experience occurs at the level of Marr's 2.5D sketch.

(1*) *The empirical premise:* Marr's theory of visual perception, or something similar enough to it, is true.

(1a*) *The Prinz addendum to the empirical premise:* Neuroscience shows that Marr's 2.5D sketch is realized in extrastriate cortex.

(2*) *The descriptive premise:* Of the three stages of visual representation in Marr's theory, the 2.5D sketch describes what we visually experience.[5]

(3*) *Conclusion:* Therefore, visual experience occurs at the level of Marr's 2.5D sketch.

Both Jackendoff and Prinz endorse the empirical premise. Jackendoff allows that Marr's theory is incomplete, but approves of Marr's "basic insight that it is crucial to ask what sort of information the visual system delivers before attending to the real-time details of how the information is processed and stored" (1989: 178). Qualifications aside, he "assume[s] that Marr's theory is approximately correct" (ibid.). Prinz goes into some detail about the ways in which the contemporary cognitive neuroscience of vision departs from Marr's theory. Among these ways, the most significant is that the processing of various features of the 2.5D sketch is distributed throughout extrastriate cortex, in areas V2, V3, V4, and V5, along with perhaps some other areas (2012: 52). In other words, there is not evidence for a particular location for the 2.5D sketch in cortex. But again, qualifications aside, Prinz follows Jackendoff's main idea:

> Despite these differences between Marr's theory and the picture that emerges from visual neuroscience, the prevailing conception of how vision works is broadly consistent with Marr on the points that matter for Jackendoff's conjecture. … More to the point, findings from neuroscience have added support to Jackendoff's conjecture that [the 2.5D sketch] is the locus of conscious experience. (2012: 54)

Prinz marshals a great deal of empirical evidence in support of this claim. I will turn to some of this evidence shortly. Before doing so, I should note that both Jackendoff and Prinz endorse the descriptive premise of the argument as well.

The reasoning behind the descriptive premise is as follows. If Marr's theory is correct, per the empirical premise, then there are three candidate forms of representation that might determine conscious visual experience. The first candidate is the primal sketch, which can be ruled out because it lacks the organization, depth, and binocular fusion of conscious visual experience (Jackendoff 1989: 293). Another candidate is the 3D model representation, which can be ruled out because, among other things, it is perspective-independent, whereas conscious visual experience is clearly perspectival (ibid.). The middle candidate, the 2.5D sketch, is just right. It possesses binocular fusion, depth, organization, and it is perspectival. Prinz explains:

Of Marr's three levels, only the 2.5D sketch corresponds to conscious experience. We consciously perceive a world of surfaces and shape oriented in specific ways at various distances from us. If Marr is right about the three levels of vision, then Jackendoff seems to be right about the stage at which visual consciousness arises. (2012: 52)

The conclusion to the Jackendoff/Prinz argument is that conscious visual experience is determined by the 2.5D sketch and located in extrastriate cortex.

Here are a number of similarities between their general argument and mine. The obvious similarity is the general strategy, which involves both empirical and descriptive claims. On the empirical side, I agree that humans are able to extract 2.5D information about the visual scene, and that humans are capable of visually recognizing objects. It would be extreme, I think, to deny this. Also, I agree that extrastriate cortex plays an important role in human visual processing—the evidence for this claim, as Prinz demonstrates, is overwhelming. On the descriptive side, I agree that conscious visual experience is perspectival. Indeed, the perspectival nature of visual experience is a central point of the following chapter.

Now here are the main differences between their argument and mine. I will cover the differences by presenting my misgivings about each of their premises in order. By doing so, I will introduce my reasons for departing from their views, reasons that will serve as the beginnings of my support for the two premises of my own Main Argument. Let us begin with the first premise of the Jackendoff/Prinz argument.

(1*) *The empirical premise:* Marr's theory of visual perception, or something similar enough to it, is true.

This premise signals a general allegiance to Marr's framework, a framework that has been both widely adopted (especially in computer vision; see Tappen, Freeman, and Adelson 2005, for an example) and heavily criticized (Churchland, Ramachandran, and Sejnowski 1994). My view is that some of the criticism is strong enough to warrant a rather significant departure from some of Marr's core ideas. In its general form, the strongest criticism is that Marr's explicit decision to abstract away from bodily details obscures the crucial fact that human vision is typically a process of active exploration. As a result, it neglects what I referred to as the "weak claim" above. The kinds of empirical evidence often cited in support of active (or enactive) approaches to vision bring out the importance of action for understanding human vision. These include selective rearing paradigms (Held and Heine 1963), perceptual adaptation (Kohler 1964), change blindness (Rensink, O'Regan, and Clark 1997), and inattentional blindness (Mack and Rock 1998). I will turn to these, and other, lines of evidence in chapter 5. The abstraction from bodily details also misses the fact that the photoreceptor distribution on the retina is not uniform (Lindsay and Norman 1977). This fact is likely one main cause of our experience of peripheral indeterminacy. The indeterminacy of visual experience, especially peripheral indeterminacy, is a main theme throughout the entire book beginning with the following chapter. Also, the details of photoreceptor distribution on the retina play a central role in my interpretation of the evidence for two visual systems in chapter 7. There, I argue for an alternative to the standard interpretation (Milner and Goodale 1995) of the dorsal stream as processing unconscious "vision for action" and the ventral stream as processing "vision for perception," which may become conscious. My alternative is that the main difference between the streams is spatiotemporal. Briefly put, the dorsal stream processes peripheral input quickly, while the ventral stream processes foveal input more slowly. In sum, my main reasons for departing from (1*) can be found in chapter 5, with implications throughout the entire book, especially chapter 7.

Now consider the next premise of the Jackendoff/Prinz argument.

(1a*) *The Prinz addendum to the empirical premise:* Neuroscience shows that Marr's 2.5D sketch is realized in extrastriate cortex.

Suspicion about (1*) will lead to suspicion about (1a*). That is, if there are good reasons for departing from Marr's framework, then there are good reasons not to impose that same framework onto our interpretation of neuroscientific evidence. While the neuroscientific support for premise (2) of my Main Argument can be found mostly in chapters 6 and 7, I should address several issues here as a way of setting the stage. In what follows, I will cover three general reasons to reject (1a*), and offer some general remarks against anatomical localization. Due to the complexity of the topic and our current level of knowledge, I think it would be premature to claim that these reasons are decisive. The fact is that there is still very much we do not understand about the brain. With this caveat in place, my discussion here is intended to offer strong empirical reasons to be skeptical of (1a*), and thus to consider an alternative, in particular, to consider my premise (2). My impression is that the understanding of brain function at use in Prinz's framework is one that is orthodox within cognitive neuroscience and rather widely accepted within empirically oriented philosophy of mind (see Block 2007 for another example). Importantly, though, there is a good amount of uncertainty with regard to what is the "orthodox" or the "received" view in these matters. There are many researchers working on this topic, and the opinions of the researchers themselves are always changing. Readers should keep in mind that my criticisms of the historically orthodox view in cognitive neuroscience may, at times, target views that are already on the wane. I have tried to paint an honest picture of the major theoretical camps in cognitive neuroscience, but with wide strokes for the sake of keeping the dialectics straightforward. The historically orthodox view in cognitive neuroscience is, I take it, nicely characterized by the way in which Prinz supports (1a*). Later on in the book I will be appealing to an alternative way of understanding the functioning brain that will be new to some readers and marks a departure from the historically orthodox position. My critical discussion of (1a*) is a way of introducing reasons to begin considering the alternative.

Prinz's methodology in bringing together psychology and neuroscience is a version of what Francis Egan and Robert Matthews (2006) have called *top-down neuroscience*. According to Egan and Matthews, top-down neuroscience "presumes that one should begin with a well-developed psychological theory, perhaps computational in nature, and then look for the manner in which the various states and processes postulated by the theory are implemented" (2006: 379). They go on to point out that—no surprise—David Marr (1983) is one of the best known advocates of such an approach. The approach is top-down because it begins with a particular model of some faculty of mind and then tries to map that model onto brain structure. In particular, the model is one in which the task, in this case visual perception, is broken down into subtasks, which can then be broken down into still more basic subtasks. Readers will be familiar with this strategy as homuncular decomposition (Dennett 1987). Prinz adopts this strategy:

> Computational devices decompose into various interconnected subsystems, each of which performs some aspect of a complex task. Given such a decompositional analysis, we can ask, in which subsystems does consciousness arise? (2012: 49)

In addition to functional decomposition, Prinz also adopts a strategy of anatomical localization. As (1a*) makes clear, he thinks that we can pinpoint the particular anatomical region in which visual consciousness occurs in the form of Marr's 2.5D sketch. It is reasonable, perhaps conceptually true, that a functional-level description of a computational process admits of this kind of decomposition. But there are a number of reasons to be skeptical of a strategy in which what goes on at particular locations in cortex is interpreted through the lens of functional decomposition. Such a strategy, at least as deployed by Prinz, involves mapping subsystems onto particular structures in cortex.[6]

I will begin with three reasons to reject claims of a mapping from functional subsystems onto localized structures in cortex, the kind of mapping central to Prinz's support for (1a*). After covering these three reasons, I will offer some further critical remarks about the general strategy of looking for the precise location of visual consciousness in the brain. The three reasons are as follows:

1. Cortical areas typically do not have only one particular function.
2. Neuronal response appears to be context-sensitive.
3. The small-world connectivity of cortex does not support anatomical localization of function.

These reasons, in one form or another, can all be found in Egan and Matthews (2006: 385), but I have added further elaboration of all three. Considered together, these reasons reflect a general tension in mapping relatively static cognitive structures onto an inherently dynamic neural substrate.[7]

The first reason for skepticism about (1a*) is that cortical areas usually do not have only one particular function. Instead, particular regions appear to be involved in a number of different tasks. Egan and Matthews cite evidence about Broca's area in support of this claim (Hamzei et al. 2003, Manjaly et al. 2003). More recently, Michael Anderson (2007, 2010, 2014) has amassed a great deal of evidence for this claim in his development of the neural reuse hypothesis. One especially convincing line of evidence stems from Anderson's use of the Neuroimage-based Coactivation Matrix, or NICAM, database of over 2,000 functional magnetic resonance imaging (fMRI) studies. He considered 11 different task domains, composed of two perceptual (vision and audition), three action (execution, inhibition, and observation), and six cognitive (attention, emotion, language, mathematics, memory, and reasoning) domains. Then he chose a spatial subdivision of the brain into 986 regions of interest (following Hagmann et al. 2008), 968 of which were activated in at least one of the studies. Anderson found that 91.8% of those 968 regions were active in at least two of the task domains. He also found that the 968 regions of interest were active in an average of 4.32 task domains. (See Anderson 2010: 258 and Anderson, Brumbaugh, and Suben 2010 for further details.)

A second reason to reject the mapping between functional subsystem and local areas of cortex is that brain function tends to be context-sensitive, depending, for instance, on attention, task, and the immediate history of sensory stimulation (Egan and Matthews 2006: 385). The context sensitivity of neural processing speaks against one of Prinz's main sources of evidence for (1a*), namely evidence from single-cell recordings in nonhuman primates. Prinz points out that the firing response of extrastriate cells correlates well with conscious visual experience, especially if contrasted with cells in neighboring areas of visual processing such as V1 or IT (2012: 55). This fact alone, however, cannot "be used to determine what kind of information is encoded in different stages of the visual hierarchy" (2012: 54). It cannot be used in this way because the firing response of particular extrastriate neurons appears to be context-sensitive. If the way in which neurons respond depends on various facts about the organism's context, then it would be a mistake to draw wide-ranging conclusions based only on the particular context of single-cell recordings in the lab, a context in which, in Michael Spivey's memorable phrase, "the animal is slowly dying on a slab with its eyelids propped open" (2007: 209).

Spivey emphasizes that the neural response to visual stimuli can depend on the previous actions of the organism; for instance, eye movements alter the neural response to the stimulus (see Spivey 2007: chapter 8 and the many references therein, especially Duhamel, Colby, and Goldberg 1992). Evan Thompson and Alva Noë (2004) first made the point about the context-sensitivity of neuronal response as a general objection to the practice of using the firing response of a neuron as justification for claims about the content of that neuron matching the content of conscious experience. They cite evidence that the recorded response of neurons in visual areas is sensitive to an animal's bodily position (Horn and Hill 1969, Abeles 1982, Abeles and Prut 1996), to auditory stimulation (Fishman and Michael 1973), and to the task with which the animal is engaged (Haenny, Maunsell, and Schiller 1988, Treue and Maunsell 1996). I myself have taken up this theme by focusing on evidence for context sensitivity in the response of neurons in area V5/MT (Madary 2013b), which is a part of extrastriate cortex, Prinz's purported location of visual consciousness. Maier, Logothetis, and Leopold (2007) recorded from V5/MT in a binocular rivalry paradigm and found changes in the functional role played by particular neurons across subtle changes in the stimulus. Similarly, Cohen and Newsome (2008) recorded from V5/MT neurons in monkeys performing a directional discrimination task. They changed the perceptual context by changing the direction of movement that the monkeys were supposed to discriminate—same stimulus, slightly different task. They found that just this change in context resulted in a change in the functional circuitry of neurons within V5/MT. The recorded response of extrastriate neurons turns out to be much more puzzling than top-down neuroscience might lead one to believe.

A last reason from Egan and Matthews to reject a mapping from a functional subsystem onto particular areas of cortex is due to the nature of the connectivity in cortex. In chapter 6, I will discuss the massive feedback connectivity in cortex. For now, I will mention another property of neural connectivity that does not fit well with (1a*). That property is the mixture of short-and long-range connections for typical cortical neurons. Although the details of neural connectivity are still being explored—in the Human Connectome Project,[8] for instance—there are some properties that are fairly uncontroversial. There is a general consensus that, by estimation, many thousands of cortical neurons form together into local circuits, and that *each* neuron in those circuits forms tens of thousands of synapses with other neurons (Braitenberg and Schüz 1998: 190, Binzegger, Douglas, and Martin 2004, Bock et al. 2011). Most of these connections are thought to be relatively local, but, importantly, recent work has discovered a surprisingly large fraction of long-range connections (Stepanyants et al. 2009).[9] Apart from the recent findings, there are also well-known long-range connections between distant areas of cortex, classically known as association fibers. For example, one such fiber, known as the occipitofrontal fasciculus, connects the frontal lobe with the occipital lobe. This pattern of connectivity creates what is known as a "small world" network (Watts and Strogatz 1998), a network in which all neuronal elements are connected to all other neuronal elements in the cortex by only a few connections—despite the fact that the majority of connections are local (Sporns and Zwi 2004, Bassett and Bullmore 2006). Although neither neural connectivity, nor the functional implications of that connectivity, are fully understood, my purpose in mentioning these features is to raise the following point: the small-world connectivity of cortex is *prima facie* not the kind of connectivity that one would expect if Prinz's (and Marr's) top-down approach to visual neuroscience is correct. On their view, processing tasks are broken down into simpler and simpler further tasks. It is reasonable to think, following Fodor, that these simpler tasks must be completed with some degree of information encapsulation. But small-world neural connectivity does not reveal a pattern of information encapsulation; instead, it shows a delicate balance between information encapsulation and some kind of information promiscuity. Small-world networks enable the brain to exhibit complex, large-scale dynamics (Sporns 2011: 125), features that are not what one would expect to see in the implementation of the fixed decomposable computations posited by top-down neuroscience.

Before moving on to the final premise in the Jackendoff/Prinz argument and the conclusion of the chapter, I should make a few further remarks about (1a*). In particular, I want to express my reservations regarding the strategy of hunting for the particular place in which visual consciousness occurs in the brain. I do so because this search for the location of visual consciousness has been a major topic in consciousness studies, and I would like to justify my reasons for avoiding it, for investigating the general structure of visual processing instead of the precise location. The first reason is taken from Dennett and Hurley's work on the topic, and the other reasons have to do with fMRI and cases of brain damage.

Dennett and Hurley have both argued that there are good empirical reasons to consider the various temporal and spatial scales of perceptual processing. The importance of temporal scales for perceptual processing was famously emphasized by Dennett in his attack on the "Cartesian theater." Using results from visual masking experiments, Dennett made the case that there is no such theater, that there is no place in which "it all comes together" in the brain for visual consciousness (1991: 107). Hurley was influenced by Dennett here, and extended the point to include spatial scales as well: "Temporal extension leads to spatial extension" (2010: 111). Given the importance of bodily action for visual perception (Hurley 1998; also section 1.2 above and chapter 5), Hurley saw no reason to draw the relevant boundary around the skull for the physical basis of visual experience.[10]

My own approach to the topic has been influenced by Dennett and Hurley. The neural and bodily implementation of the cycle of anticipation and fulfillment strikes me as being distributed both spatially and temporally in a way that makes it difficult to draw boundaries. Efforts to draw those boundaries outside of the skin and skull (Menary 2010) have notoriously run into the purported causation/constitution fallacy. The objection is as follows: we can all agree that environmental and bodily processes play a *causal* role in visual experience, but it would be a mistake to then conclude that they *constitute* visual experience (Block 2005, Prinz 2006b, Adams and Aizawa 2008). Opponents of the extended mind hypothesis maintain that parts of the brain constitute mental states, while extra-neural processes merely play a causal role.

Here I hope to avoid the debate altogether by following Don Ross and James Ladyman's take on the distinction (2010). They have argued that the distinction itself is one from analytic metaphysics, and, with the help of some examples, they claim that it has no place in the mature sciences. But it appears to me that one must accept the distinction as being important in cognitive neuroscience in order even to enter the debate over pinpointing the location of visual experience. As I am convinced by Ross and Ladyman's claim that the distinction does not have a place in the mature sciences, I will opt out of the debate.

Besides conceptual worries, I suggest that there are two other reasons to be wary of anatomical localization claims in the brain. The first worry has to do with the overambitious interpretation of fMRI results. Anderson's work gives us good reasons to be wary of common interpretations of fMRI data, according to which activation in a particular region during a particular task is taken to mean that the region is uniquely dedicated to processing that task. This point is relevant here because Prinz appeals to a number of fMRI studies as a part of his support for (1a*), and brain imaging is a central part of mainstream cognitive neuroscience these days, as a quick look at popular science media will reveal. Here I would like to introduce some additional reasons for my skepticism about using fMRI to support (1a*) in particular, and top-down cognitive neuroscience in general. First, though, I should mention Prinz's reasons for citing fMRI data in support of (1a*). Namely, there is a great deal of evidence indicating that the activation detected in extrastriate areas correlate with consciously perceived visual stimuli. Prinz cites evidence for extrastriate activation for perception of stereoscopic depth (Backus et al. 2001), bi-stable figures (Lumer and Rees 1999), and viewpoint sensitivity (Konen and Kastner 2008), among other features.

Apart from Anderson's work, additional reasons for skepticism about using the fMRI results to support (1a*) can be seen if we consider some facts about what fMRI actually measures. Sometimes the literature reporting fMRI results can be misleading, such as when there are reports of unqualified activation in a particular area, reporting simply that the area "lights up." Such language is misleading because it can give one the impression that these areas are otherwise deactivated, as if they are just waiting around for the right stimulus or task to light them up. The fact is that the entire brain is always "active," using a disproportionate amount of the body's metabolic energy relative to the size of the brain. The activations discovered using fMRI are typically very small increases in metabolism—less than 5%—as measured by changes in oxygen levels from the blood. In an overview of the state of the art with regard to fMRI, Nikos Logothetis concludes:

The fMRI signal cannot easily differentiate between function-specific processing and neuromodulation or between bottom-up and top-down signals, and it may potentially confuse excitation and inhibition. Further, the magnitude of the fMRI signal cannot be quantified to accurately reflect differences among brain regions or among tasks within the same region. (2008: 877)

Brain-imaging pioneer Robert Shulman has drawn even stronger conclusions based on the sort of considerations mentioned above. He has made the case that fMRI cannot be used to support top-down neuroscience, or what he calls the "computer model of the brain" (2013; for a review written with philosophers in mind, see Klein 2013).

Another line of empirical evidence used to justify (1a*) comes from cases of brain damage, in particular cases of damage to extrastriate cortex. Prinz (2012: 56–57) mentions evidence that damage to area V1, "downstream" from extrastriate, as well as damage to extrastriate itself, is sufficient to cause blindness. He also mentions that damage to visual areas "upstream" from extrastriate do not cause blindness. This kind of evidence, in itself, only suggests that damage to these particular areas is sufficient, in the case of V1 and extrastriate, or insufficient, in the case of the "upstream" areas, to cause blindness. In general, when lesions cause the loss of an ability, it may not be right to conclude that the lesioned area is the location of the processing for that ability (van Orden, Pennington, and Stone 2001, Davies 2010). One main reason to resist this conclusion are cases of the "Sprague effect," in which a lesion causes the loss of an ability and that ability is then *restored* by a second lesion (Sprague 1966, Weddel 2004). A second reason to be skeptical is that neural network models have been constructed in which random damage can knock out a particular ability, despite the fact that there is not localized processing for that ability in the network (Juola and Plunkett 2000, Plunkett and Bandelow 2006). In any case, the evidence does not reveal that there is normally a representation of a particular kind within the damaged area, or that the damaged area is dedicated to a particular functional role. In order to make that further inference, we should have independent evidence for anatomical localization in the brain.

None of this, I should add, is decisive against premise (1a*). It is reasonable to think that there is *some* form of anatomical localization in the brain; Prinz's various lines of evidence indicate as much. It might turn out that extrastriate cortex is put to use during a number of distinct tasks, but that generating the 2.5D visual representation is a task that occurs strictly in the extrastriate cortex. And it might turn out that neural network models with distributed processing channels are functionally dissimilar to our brains. But I raise these objections in order to show that the evidence we have so far does not support strong confidence in claims about anatomical localization currently on offer. We need an understanding of the brain that does not ignore basic features of the cortex for the purpose of oversimplifying the relationship between mental function and neural substrate.

My strategy is not one of localization, nor is it top-down in the sense of Egan and Matthews. Instead, it is a methodology that takes seriously some properties of the brain itself, including those listed above, without prior constraints or presuppositions from psychology. In particular, my view is very much influenced by the fact that the brain has its own ongoing endogenous dynamics (see section 6.1 in this volume; Buzsaki 2006, Bechtel 2013). To be clear, I am not suggesting that anyone denies that the brain has its own ongoing endogenous activity. But I am suggesting that this activity has not always been given great importance for understanding brain function. As I explain in more detail in chapter 6, the ongoing dynamics of cortex suggest that the brain is an *active*, rather than a *reactive*, organ (Clark 2015b). Understanding the brain as a reactive organ fits nicely with Prinz's appropriation of Marr's framework: the retinas receive input, and processing proceeds hierarchically from there. In contrast, taking the brain to be an active organ, with its own intrinsic dynamics, fits better with my premise (2), according to which the brain spontaneously generates anticipations. Taking the brain to be an active organ also fits well with the explosion of recent work on predictive processing (Clark 2013b, 2015a, Hohwy 2013), much of which I take to provide strong support for (2) and revisit in chapters 5 and 6.

Finally, consider premise (2*) of the Jackendoff/Prinz argument.

(2*) *The descriptive premise:* Of the three stages of visual representation in Marr's theory, the 2.5D sketch describes what we visually experience.

As indicated above, I incorporate a central part of this premise in my premise (1). The strong appeal of Marr's 2.5D sketch is that it captures the perspectival nature of perception. As readers will see in the following chapter, the perspectival nature of experience is central to my thinking of vision in terms of anticipation and fulfillment. But there are two other constraints that I introduce in the next chapter that are not included in (2*). These are the temporal and the indeterminate nature of visual experience. The 2.5D sketch is a static image, and it has already become clear in vision science that it is not amenable to the problem of how we experience the visual world over time. That is, the 2.5D sketch alone cannot accommodate the *integration* of different viewpoints, known as trans-saccadic perception (Wexler 2005, Melcher and Colby 2008, Prime, Vesia, and Crawford 2011). Similarly, because the 2.5D sketch leaves out the details of biological implementation, it does not include peripheral indeterminacy. In the following chapter I will introduce all three constraints on visual phenomenology, and I suggest that (1) meets all three constraints. Indeed, (1) is deliberately formulated as a way of meeting all three constraints. In the third chapter, I offer an argument for (1) itself in the form of a *reductio ad absurdum*. There I also introduce five main features of visual anticipation. Two important such features are, first, that visual anticipations have various degrees of determinacy and, second, that they are ongoing, rather than episodic.

The fourth chapter of the book then covers the question of visual content. There I urge a departure from the mainstream practice in the philosophy of perception of understanding visual content as propositional. I suggest that this departure solves some outstanding puzzles in the philosophy of perception: the speckled hen (Nanay 2009), skepticism about introspection (Schwitzgebel 2008), and phenomenonal overflow (Block 2007).

To sum up, I have followed the same general strategy of the Jackendoff/Prinz argument, and here I have offered my main reasons for departing from some of the details of their argument. My appropriation of their strategy should be clear in the structure of my Main Argument:

(1) *The descriptive premise:* The phenomenology of vision is best described as an ongoing process of anticipation and fulfillment.
(2) *The empirical premise:* There are strong empirical reasons to model vision using the general form of anticipation and fulfillment.
(AF) *Conclusion:* Visual perception is an ongoing process of anticipation and fulfillment.

In comparing my argument with the Jackendoff/Prinz argument, readers may notice that the order of the premises is switched around. This is merely a stylistic decision on my part. Unlike the Jackendoff/Prinz argument, my premises are independent of each other, and I will begin with the phenomenology before moving on to the empirical results.

2 Three Constraints

In this chapter I introduce three general features of all visual experience: visual experience is perspectival, temporal, and indeterminate. Any account of visual phenomenology ought to accommodate these three constraints. Here I am not assuming any particular methodology about investigating visual phenomenology, except that experience can and should be investigated, with care, from the first-person perspective (Siewert 1998). The three constraints are intended to be completely uncontroversial in this regard.

I have claimed that my support for (1) consists in describing visual experience, and I should begin by making a quick remark as to what I mean by that claim. As readers will see, I will be indicating general features of visual experience that we can all observe. By calling them descriptive, I mean that these features can be discovered without experimental investigation. That is, my case for (1) does not depend on results from psychology or neuroscience. Again, I do not mean to imply that (1) is a necessary truth (see the appendix, section A2, for a discussion of this issue).

As we began to see in the previous chapter, Marr's 2.5D sketch is sensitive to the perspectival nature of visual experience, but it neglects the temporality and indeterminacy constraints. It is tempting to neglect the temporality and indeterminacy of perception if one conceives of visual experience as a snapshot image taken with a camera. As Larry Hardin (1988: 7–18) and, more recently, Noë (2004: chapter 2) have argued, there are good reasons for rejecting this conception of vision. My justification for temporality and indeterminacy will adopt some of these reasons and introduce some new ones. In what follows, I make the case for each constraint in turn, beginning with perspective.

2.1 Visual Experience Is Perspectival

The perspectival nature of visual experience has been a theme in the philosophical literature at least since early modernity, having been discussed by Leibniz (Swoyer 1995) and Hume. Hume illustrates the perspectival nature of perception using the example of a table: "The table, which we see, seems to diminish, as we remove farther from it" (1777/1993: 104). Since then, philosophers have appealed to a number of different examples in order to illustrate the perspectival nature of vision, including Husserl's red sphere (1900/1993: V §2), Moore's envelope (1953: 30), Russell's table (1912/1978: 2–3), Ayer's tilted coin (1936/1952: 67), and Peacocke's trees (1983: 12).

The perspectival nature of visual experience generates a puzzle. On one hand, our experience is limited to one perspective at a time. On the other hand, most of the properties we deal with and talk about are properties that can be experienced from multiple perspectives. In discussing this puzzle, it may be helpful to introduce a distinction between properties that depend on particular perspectives and properties that do not depend on perspective. This distinction is familiar within recent philosophy of perception, in which one can find discussions of perspectival versus factual properties (Noë 2004), situation-dependent versus intrinsic properties (Schellenberg 2008), and viewpoint-dependent versus viewpoint-independent properties (Brogaard 2010). I will adopt Noë's terms, since he seems to have been the first recently to focus on the distinction:

Perspectival properties are properties that can only be perceived from a particular perspective.
Factual properties are properties that can, in principle, be perceived from many perspectives.

There are some differences about the details with which each author makes the distinction, but I do not think these details are necessary to include here. The important point is that this distinction helps us to explore some of the difficulties that arise in considering the perspectival nature of perception. A main advantage of (1), I suggest, is that it offers a nice way to overcome these difficulties; it offers an account of how we perceive factual properties without relying on the somewhat controversial claim that we represent perspectival properties.

There are a number of philosophical issues surrounding the relationship between perspectival and factual properties. Schellenberg (2007) has suggested, for instance, that one's representation of factual properties depends epistemically on an accurate representation of perspectival properties. Noë (2004, 2012) has suggested that we visually experience both perspectival and factual properties simultaneously, while others have been skeptical about our ability to experience perspectival properties (Smith 2000, Kelly 2004, Siewert 2006, Briscoe 2008, Overgaard 2010, Hopp 2013, and perhaps Schwitzgebel 2006). Whether or not we represent perspectival properties, a puzzle remains about how we are able to represent factual properties despite always only having partial, or perspectival, access to them.[1] As I explain in section 2.4 below, (1) gives us an elegant way to deal with this puzzle.

Before moving on to the temporality constraint on visual experience, I should explain my own view on perspectival properties. Thesis AF does not involve any representations of perspectival properties, which makes it unlike the approaches of Schellenberg and Noë. This difference is an advantage because there is no evidence that humans are any good at tracking perspectival properties. I suppose one could posit a subpersonal representation of perspectival properties for theoretical reasons; one could posit a representation which would not be accessible for the standard paradigms used in psychophysics measurements. Also, one could include orientation relative to one's body as a perspectival property, which is something that we obviously are able to track. But I am aware of no evidence through conscious reportability that humans accurately represent other perspectival properties such as perspectival size and color.[2] Here is a description of some of the relevant experiments, in a brief violation of my claim that I save the empirical evidence for later chapters.

Lawson, Bertamini, and Liu (2007) ran a series of experiments designed to test whether humans track the projection plane account of perspectival size, as calculated using visual angle, in perceiving everyday objects. In one experiment, normal subjects stood before a window through which they were able to see a bamboo stick. They were told to use a measuring tape held at their waist to estimate the size of the stick. The subjects were also told to estimate the projected size of the stick on the window. In order to prevent subjects from gaining an advantage by looking at the reflection of their arm, subjects were instructed to pull the tape measure perpendicular to the window.

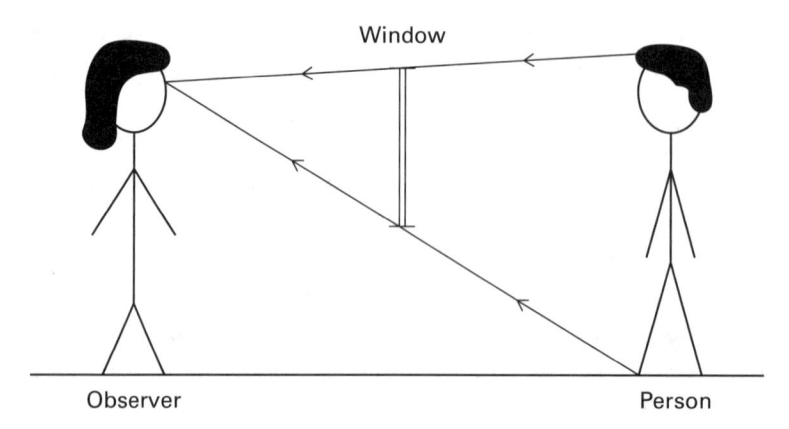

Figure 2.1
Calculation of perspectival size using a trigonometric method. The height of the person relative to the observer and the window is exactly the height of the window. Reprinted with kind permission from Lawson et al. (2007).

Projected size was calculated using simple trigonometry, as in figure 2.1. Subjects were presented with a number of sticks, between 10 and 80 cm, and care was taken to ensure that subjects understood what they were supposed to estimate as "projection size on the window." The results were clear. Subjects could accurately estimate the actual size of the bamboo sticks, but consistently overestimated the projection size. According to the trigonometric calculation of projection size, the subjects should have reported projection size to be half of the actual size of the stick. Instead, most reports of the projection size of the stick were closer to 80% of the actual size of the stick. Lawson, Bertamini, and Liu suggest that "there is no percept for the 2-D projection on the surface of a mirror or a window" (2007: 1028). The poor performance in the projection estimation task cannot be attributed to the fact that subjects were reporting with a perpendicular tape measure; they used the same method of reporting for the size-estimation task, which they performed quite well. In these experiments, we have a direct test of whether or not normal subjects accurately track the property of perspectival size as described by Tye (2000: 79) and Noë (2004: 84). The results show no evidence that we track this property.[3]

In addition to perspectival size and perspectival shape, there is also what we might call perspectival color.[4] Take a uniformly colored object, say a red sphere. Then look at it under a number of different lighting conditions:

one might want to deny my temporality constraint due to a Dennettian flavored interpretation of some empirical results. By appealing to a range of surprising psychophysical results, Dennett has proposed his "multiple-drafts" model of conscious experience. The model suggests that there is no fact of the matter about the content of our conscious experience at any particular time (1991: chapters 5 and 6). As my discussion in chapter 4 will make clear, I am generally sympathetic with this way of thinking about perceptual content, and see no conflict between it and the temporality constraint (following Hurley 1998: 51). That is, I allow that the content of visual experience can be messy and indeterminate, while maintaining the widely held view that the temporal flow appears to be continuous. Even if there is some grand illusion about the content of visual experience (Noë 2002), we need not draw the even more radical conclusion that there is an illusion about the way experienced time appears to us. Even Dennett himself does not deny that there is a stream of consciousness, using the term throughout *Consciousness Explained*, as in "there are no fixed facts about the *stream of consciousness* independent of particular probes" (1991: 138, emphasis added). I will return to similar issues in section 4.4. and in my discussion of change blindness in section 5.2. Another empirical phenomenon that might raise suspicion about the temporality constraint is akinetopsia, which I will cover in section 7.5.

A second way to resist the temporality constraint would be to claim that the apparent continuity of experience over time is not important for perceiving the world. One might say, for instance, that object perception does not depend on the seamless integration of different viewpoints. Susanna Schellenberg has developed such a position in her discussion of the *unification problem* of perception, which is the problem of how different viewpoints are "integrated into the perception of an object" (2007: 609–610). Schellenberg attempts to sidestep the problem by suggesting that "perception of intrinsic spatial properties does not depend on subjects having two encounters with an object (either past or present). Just one encounter is required" (2007: 618). If we could sidestep the *unification problem*, then perhaps we could also safely ignore the temporality of experience. Such a move strikes me as unpromising, however, because Schellenberg's strategy for avoiding the *unification problem* is open to counterexamples.

The problem with Schellenberg's way of sidestepping the *unification problem* is that, as Husserl puts it, "it may be that a 'single glance' is not important point, as I explain below, is that AF reflects the temporal nature of perception in a straightforward manner. The dynamical nature of visual experience is built in, as it were, into the structure of the experience itself.

The natural fit between AF and temporality marks another advantage of (1) over some ways in which visual experience is commonly approached in the philosophical literature. In chapter 4, I will discuss some of the ways in which perceptual content is usually treated and offer reasons for departing from some of these ways. Here I only note that the temporality of visual perception is a general feature, and that it would be better to adopt a theory of visual perception that is sensitive to that fact rather than in tension with it. Historically, the methods of perceptual psychology have involved presenting static images as visual stimuli for subjects. This way of investigating vision may be a main reason why temporality has been somewhat neglected in psychology and empirically informed philosophy of mind as an essential feature of visual perception. Here I urge that the technological limitations in the perceptual psychology laboratory should not overly influence our understanding of vision. With the increasing availability of virtual reality for psychological research (Madary and Metzinger 2016), the temporal aspects of vision will soon become more easily approached in the lab.

The temporality constraint is meant to express the simple claim that visual experience appears to be *continuous* over time. We do not experience the visual world as a series of discrete snapshots. It is this feature of consciousness that William James famously described as a "stream":

Consciousness, then, does not appear to itself chopped up in bits. Such words as "chain" or "train" do not describe it fitly as it presents itself in the first instance. It is nothing jointed; it flows. A "river" or a "stream" are the metaphors by which it is most naturally described. (1918, vol. I: 239)

I agree with James that the metaphor of the stream is a natural way to describe the temporality of experience (also see Siewert 1998: 349n4), while leaving open the philosophical fine points. Arguably, his term endures in popularity because people find it to be an effective way to describe the continuous flow of time.

Just to avoid misunderstanding here, the temporality constraint is a claim about the *appearance* of visual consciousness. As such, I think it should be uncontroversial. I intend it to be compatible even with skepticism about our ability to introspect our own conscious states. For instance,

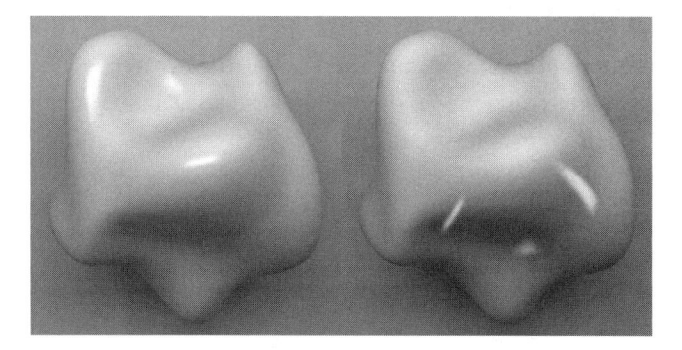

Figure 2.2
The object on the left looks glossy because the specular highlights are in the "right" places. The object on the right has highlights in the "wrong" places, which makes it look like a matte surface with spots of white paint. Image reprinted with kind permission from Todd, Norman, and Mingolla (2004).

not accurately represent the location of specular highlights relative to surfaces (for more details, see Madary 2008).

The important point of this section is that visual experience is perspectival. Much of the philosophy of perception has involved attempts to deal with the relationship between perspectival appearances, on one hand, and the representation of factual properties, on the other. As I will argue below, premise (1) offers a nice way of accounting for this relationship. Premise (1) has a further advantage over a number of contemporary approaches in that it does not depend on the empirically suspicious assumption that we normally represent perspectival properties in experience. This second advantage of premise (1) will resurface in my discussion of some competing views in the existing literature (see sections 4.3 and 8.3).

2.2 Visual Experience Is Temporal

Just as Leibniz and Hume were sensitive to the perspectival nature of perception, so Kant was famously concerned with the temporal nature of all experience, as he emphasized in the second section of the Transcendental Aesthetic: "Time is a necessary representation (*Vorstellung*), which lays at the foundation of all intuitions" (1781/1998: A31). We experience the world in time. The details of the relationship between perception and temporality are notoriously difficult, and an issue that I will not cover here.[5] But the direct sunlight, indoor ambient light, under a lamp. For the most part, the red sphere will continue to look uniformly red under the different lighting conditions. But there is also the sense in which the appearance of the sphere changes. The appearance can change even within the same lighting condition; for example, the location of the highlights and shadows on the surface of the sphere might change as you take different perspectives on it. To apply our terms from above, the uniform redness is the factual color of the sphere and the changing surface appearance is the perspectival color of the sphere.

In a well-known study from 1986, Lawrence Arend and Adam Reeves set out to test whether subjects are able to distinguish between factual color and perspectival color (to use our terms). They presented subjects with a colored stimulus under various lighting conditions and instructed them to adjust another test square in order to have it match the stimulus. They gave subjects different instructions in order to have them match either the factual color of the stimulus or the perspectival color of the stimulus. In the condition testing for factual color match, they instructed subjects to adjust the test square to "look as if it were cut from the same piece of paper" as the stimulus (1986: 1745). In the condition testing for perspectival color match, they instructed subjects to match the "hue and saturation" of the stimulus patch (ibid.). Results indicate that subjects showed a good ability to match for factual color, for the "paper" condition, but performance was poor for the perspectival color condition. Further evidence for these results can be found in subsequent studies (Arend et al. 1991, Troost and de Weert 1991, Cornelissen and Brenner 1995).

Another candidate for a perspectival property that we might track are specular highlights, which are the perspective-dependent shiny spots appearing on glossy surfaces (see figure 2.2). Andrew Blake and Heinrich Bülthoff (1990) ran a series of experiments in which subjects were told to adjust the disparities between the input to each eye until the specular highlights appeared as if they were in the correct, or realistic, locations. Subjects adjusted the locations of the highlights to be slightly in front of or behind surfaces. Despite placing the highlights off of the surfaces, subjects reported the appearance of the highlights to be on the surfaces. Blake and Bülthoff conclude that we have implicit expectations about how highlights should appear (a conclusion that fits nicely with AF), but that our actual experience of the locations of the highlights themselves is not accurate. We do good enough" (1900/1993 VI §47). Say we are playing a game with a set of dominoes in which the pips are carved into the face of each domino. Thus, the 3D shape of each domino fully determines its value in the game. Say I deal you a domino face down. You have a single perceptual encounter with it, but still do not know its value. Since you do not know its value, you have not perceived its 3D shape from that single encounter. In order to see the shape, and therefore the value, of this face-down domino, you must turn it over. In turning the domino, you generate the *unification problem*, which is "the question of how the appearances are integrated into the perception of an object" (Schellenberg 2007: 609–610). By reflecting on the domino scenario, it should become clear that you are *able* to integrate the appearance of the face-down domino with the appearance of the face of the domino. It should also become clear that this integration is crucial for you to perceive that the domino you see face-down is the same domino you see face-up. Finally, further reflection should make it clear that these kinds of perceptual situations are not terribly uncommon.

The temporality constraint may be in tension with the claim considered in the previous chapter that we only see the facing surfaces of objects, with premise (2*) in the Jackendoff/Prinz argument. In addition to seeing surfaces of objects, the temporality constraint reminds us that we also see the way the appearances of those surfaces change over time as we (or they) move. Proponents of (2*) must either qualify (2*) in order to allow for the experience of change over time, or they must deny that we do experience change over time. The former strategy involves a substantial change to the argument, and the latter strategy strikes me as extreme and ad hoc. As mentioned in the previous chapter, the problem of unifying appearances over time is now recognized as an important open question in the science of vision, a question that Marr's framework is not well equipped to address (Findlay and Gilchrist 2003: chapter 9). The shortcoming of the traditional emphasis on single fixations is that it generates the problem of how those single fixations become integrated in order to enable the perception of a stable scene (Schellenberg's unification problem). In the recent literature, a way of dealing with this problem is to investigate neural mechanisms for integrating information across saccades, known as trans-saccadic memory (Prime, Vesia, and Crawford 2011). There is evidence that the intention to perform a saccade is accompanied by changes in neural processing both before and after the saccade itself (Melcher 2007, Melcher and Colby

2008), with additional evidence that pre-saccadic processing is anticipatory in nature (Wexler 2005). In short, focusing on single fixations, as many philosophers continue to do, generates an explanatory puzzle. The way to move beyond this puzzle is to take seriously temporally extended transsaccadic visual processing.

2.3 Visual Experience Is Indeterminate

The third and final constraint on visual experience is the indeterminacy of visual experience. In this section I will introduce peripheral indeterminacy. In the following chapter I cover some other ways in which visual experience can be indeterminate.

Peripheral indeterminacy likely occurs due to the fact that photoreceptor cells are not distributed evenly throughout our retinas. The center of our visual field, in which vision is clearest, corresponds to the fovea on the retina, which is a little pit in which the photoreceptor cells are densely packed (Lindsay and Norman 1977). Our visual field becomes increasingly indeterminate as we move from central vision toward the periphery. We do not usually notice this indeterminacy since we are able to move our eyes and bring areas of interest into the center of the visual field. With careful attention, though, we can notice peripheral indeterminacy, which has long been an important theme in experimental psychology (James 1918, Vol. II: 161–162). In the philosophical literature, both Husserl and Dennett have made this point.

In notes to lectures from 1907, Husserl wrote:

The region of clearest vision is so small and the clarity shades off so quickly, that, in general, every image actually extending beyond this smallest region will undergo changes in clarity in the case of movement, and so all the appearances, as they progress, will become richer in explication. (Husserl 1973b: 340, Rojcewicz trans. 1997: 294)

Toward the end of that century, Dennett makes the same point:

Take a deck of playing cards and remove a card face down, so that you do not yet know which it is. Hold it out at the left or right periphery of your visual field and turn its face to you, being careful to keep looking straight ahead (pick a target spot and keep looking right at it). You will find that you cannot tell even if it is red or black or a face card. ... Now start moving the card toward the center of your visual field, again being careful not to shift your gaze. At what point can you identify the color? At what point the suit and number? (1991: 53–54).

Dennett's method is effective for generating a first-hand experience of peripheral indeterminacy. Recently, there are more precise methods of investigating peripheral indeterminacy in the laboratory.

The basic idea behind Dennett's card trick has been implemented experimentally. Jeremy Freeman and Eero Simoncelli (2011) created images that are different yet appear indistinguishable when subjects focus on a particular part of the image (see figure 2.3). They have developed techniques for determining how much the periphery of an image can be altered before subjects are able to detect a change. As one can see from the sample images, the periphery can be distorted quite a bit without losing the indistinguishability between the images.

The empirical evidence obtained with images such as those in figure 2.3 makes peripheral indeterminacy difficult to deny. The degree to which the images are distorted is striking. One way to deny peripheral indeterminacy would be to claim that the entire visual field really is experienced in a determinate manner, but that a good bit of the determinate details we represent are actually misrepresentations; the accurate details are not really available.[6] Dennett seems to advocate such a move, and appeals to the fact that naïve subjects are genuinely surprised to discover the impoverished nature of their own peripheral vision (Dennett 1991: 355, 2001). The problem

Figure 2.3
When subjects focus on the gray dot in the center, they are unable to detect a difference between these two images. Image reprinted with kind permission from Freeman and Simoncelli (2011).

with this move is that there is a better, empirically justified, explanation for the surprise. The reason why it is surprising to discover the indeterminacy in our peripheral vision is that naïve subjects are not aware of the frequency and skill with which we actively explore the environment. We think that we experience more detail than we do experience because we do not notice the fact that we access detail through saccades. Once the frequency of saccades is considered—and more details on that will come later—there is simply no reason to posit a fully determinate misrepresentation of the entire visual field. Apart from the frequency of saccades, philosophers concerned with perceptual phenomenology have long emphasized the indeterminacy of visual experience. The consensus among such thinkers, including Husserl, Merleau-Ponty, Wittgenstein, and Gurwitsch (see Thompson, Noë, and Pessoa 1999), is that visual details are experienced as being *accessible in the world*, not fully represented in the visual field experienced at any one time. Despite this consensus, I must acknowledge that there are still troubling issues about perceptual indeterminacy I must leave open. In particular, I am not committed to any specific claims about the nature of the indeterminacy itself. Is the indeterminacy best characterized as a determinable (Nanay 2012), such as "some shade of green, but I know not which," as a probability density distribution over various determinate shades of green (Madary 2012b), or as something else entirely? Without overlooking the importance (and extreme difficulty) of these issues, the main point here is that there is indeterminacy in the visual periphery.

2.4 Thesis AF and the Three Constraints

The descriptive support for thesis AF comes from the first premise of my Main Argument:

(1) The phenomenology of vision is best described as an ongoing process of anticipation and fulfillment.

One central source of support for (1) is that the structure of anticipation and fulfillment is a straightforward and efficient way to satisfy all three constraints. The perspectival constraint suggests that our visual experience of factual properties is always incomplete. The temporal constraint suggests that our visual experience always seems to occur within a temporal flow. The indeterminacy constraint suggests that there is an indeterminate element to visual experience. The main idea, then, is as follows: *We represent factual properties by implicitly anticipating how those properties will appear in the immediate future.* When things are going well, the changing appearances of those properties fulfill our anticipations, which have various degrees of determinacy. Since vision is, *de facto*, tightly connected with self-generated movements, our visual anticipations are typically anticipations of how appearances will change as we move. Such movements bring what is indeterminately experienced in the periphery into more determinacy as we gain better perspectives on objects of interest. The perspectival, temporal, and partly indeterminate features of visual experience are all accounted for in the structure of anticipation and fulfillment.

Here is a good place to add a bit of clarification to my claim that the phenomenology of vision is best described as an ongoing process of anticipation and fulfillment. The claim is meant to satisfy the three constraints set forth in this chapter, constraints that reveal essential features of visual phenomenology. In meeting these constraints, I mean to suggest that all visual perception of the external environment involves anticipation and fulfillment. That is, I am not making the weaker claim that anticipation and fulfillment sometimes accompany a basic core of visual experience. Visual fulfillment is not just a vague feeling in addition to visual perception, not merely a feeling to the effect that things are going well. Rather, I am making the stronger claim that anticipation and fulfillment characterize the structure of visual experience itself.

A good way to illustrate visual fulfillment is to revisit the perspectival nature of perception, from section 2.1 above. Our view on objects is always limited to one perspective at a time. When we gain different views on the object, the appearances of the object change. If our perceptual grasp on the object is accurate, those changing appearances *fulfill* our implicit anticipations.[7] While a number of philosophers of perception, as discussed above, have sought to treat perspectival appearances as representations of a unique sort of property, my suggestion is that we treat perspectival appearances as fulfillments of anticipations. And those anticipations always depend on our being intentionally directed toward factual properties of the environment. The following two chapters will illustrate these ideas in more detail.

Now that I have presented my basic view, I would like to address an objection that may have occurred to some readers. The objection is that my view has no way to accommodate our initial visual experience of a scene.

For instance, when one first opens one's eyes after a deep sleep, how can it be right to say that this initial act involves visual anticipation? I would like to make two points in response to this objection.

First, the very act of opening one's eyes plausibly stirs up visual anticipations. That is, when we open our eyes, we expect to see something. The nature of what we anticipate seeing might be highly indeterminate, but there are anticipations nonetheless. For instance, it is likely that we anticipate seeing a visual scene made up of distinct physical objects that are not co-located. Vestibular information might inform the visual system as to the orientation of the scene (Angelaki and Cullen 2008). There may be implicit anticipations that the scene will be at least somewhat familiar to us, and so on.

The second point is that the very act of seeing any object always immediately involves anticipations of how that object will look from other perspectives. On the view I am proposing, we do not see any external objects (phenomena such as afterimages and such will be discussed below in section 3.2) without always having implicit anticipations about how those objects will appear from other perspectives.

Critical readers will note that I have yet to produce an argument for this view that the visual perception of objects always involves anticipations that have various degrees of determinacy. So far, I have only presented three constraints, and sketched how anticipation and fulfillment is a straightforward way to satisfy these constraints. In the following chapter, I offer an argument for the claim that we do visually anticipate with some determinacy. This argument builds on results from Susanna Siegel (2010). In addition, important questions remain about the cycle of anticipation and fulfillment. I will address some of these questions in the next chapter as well.

3 Anticipation and Fulfillment

In the previous chapter, I presented three abstract features of all visual experience and suggested that a description of the structure of visual experience ought to be constrained by these features. The structure of anticipation and fulfillment satisfies these constraints in a straightforward manner. But it is not enough to satisfy these constraints. I also need to provide an argument to the effect that there is visual anticipation of various degrees of determinacy at the level of conscious experience. My discussion begins with Susanna Siegel's argument for the claim that there is some anticipation in visual experience and then proceeds to the conclusion that this anticipation has some determinacy.

3.1 (PC) and Siegel's Doll

Susanna Siegel has argued that the content of visual experience includes the following expectation:

Perspectival connectedness (PC): If S substantially changes her perspective on o, her visual phenomenology will change as a result of this change. (2010a: 179)

Due to the fact that much of analytic philosophy of perception has, following standard paradigms in perceptual psychology, focused on stationary viewing of static two-dimensional images, this claim represents a significant and relatively controversial addition to the discourse. According to Siegel, this expectation can be found at the level of visual phenomenology; it is not merely a background assumption or belief. That is, (PC) actually makes up the content of our visual phenomenology (more on this in the following chapter). Siegel leaves it as an open question "whether a conditional such as (PC) but with a more specific consequent is represented in visual experiences of object-seeing" (197). (PC), as Siegel leaves it, is silent on the nature of the change that will occur as the result of self-generated movement. Here I take up the open question and answer it in the affirmative, which means that I am arguing for the following:

Specific anticipation (SA): Visual anticipation is more specific than indicated in the consequent of (PC).

In the first part of the chapter I will defend (SA), and show how the truth of (SA) motivates a reformulation of (PC) as the following:

(PC') If S substantially changes her perspective on o, her visual phenomenology will present different views of o's factual properties.

After arguing for (PC') by *reductio ad absurdum*, I will elaborate on it by making five points, the last of which forestalls a possible problem with (PC'). In the second part of this chapter I will discuss two consequences of (PC'), consequences having to do with variation in perceptual content. In the final part of the chapter I will discuss how my results relate to some conceptual distinctions.

In what follows I will assume that Siegel is correct about (PC) being represented at the level of visual phenomenology. Her argument for (PC) uses the method of phenomenal contrast[1] in conjunction with a thought experiment. Siegel suggests that we imagine two visual experiences, a Good experience and an Odd experience. In the Good experience you see a doll sitting on a shelf. After the Good experience, something strange begins to happen:

You look back at the doll on the shelf and find that it seems to have lost its independence: it moves with movements of your head as if you were wearing a helmet with an imperceptible arm extending from the front, keeping the doll in your field of view. You hypothesize that someone has somehow attached the doll to your eyeglasses using a very thin string, without your knowing it. ... You decide to test the eyeglass hypothesis by moving your eyes without moving your head, and you find that the doll seems to move with your eyes as well. It seems to be sensitive to the slightest eye movement. (Siegel 2010a: 185)

Siegel goes on to mention that your experience of the doll continues even when you close your eyes and try to place an opaque object in front of the doll. Your experience of seeing the doll is much like experiences of "seeing stars" after standing up too quickly. At the end of this sequence of odd experiences, you then have the Odd experience during which "you are standing in exactly the position you were in when you had the first experience, facing the same shelf where the doll was previously standing" (2010a: 186). Leaving out many of the details in Siegel's discussion of this thought experiment, her conclusion is that one of the differences between the Good and the Odd experiences is that the Good experience represents (PC) with respect to the doll, whereas the Odd experience does not. The conclusion, then, is that (PC) is represented at the level of visual experience. Siegel explicitly leaves it an open question whether the consequent of (PC) has any further level of specificity. Now here are my reasons for thinking that it does have further specificity.

3.2 (PC') and Five Points about Anticipation

This section consists of two parts. First I will make the case that we do visually anticipate with varying degrees of specificity, which justifies (SA) and (PC'). In the second part of this section, I will introduce five points about visual anticipation, the last of which addresses a possible problem of (PC') inspired by some of Siegel's comments (2010a).

If we accept (PC), then we accept that perceivers anticipate *some* change in visual phenomenology as a result of self-generated movement. The question here is whether the anticipation is of change in general or of some more specific kind of change. One direct way of addressing this question, I suggest, is to consider how (PC) fits with other plausible claims about visual content.

There are three relevant positions on visual content. First, one can deny that visual experience has any content, deny that visual experience represents the world to be one way or another (Martin 2006, Brewer 2006, Travis 2004). Since (PC) attributes some minimal content to experience, it is not clear that this first view is compatible with (PC). Second, one can claim that visual phenomenology is restricted to the way things appear from one's particular perspective. This view was more popular earlier in the twentieth century with sense-datum theorists (Moore 1953, Russell 1912/1978). Since (PC) expresses counterfactual content about how things will appear apart from one's present perspective, it does not seem that (PC) would fit well with this second view either. Thus, my focus will be on the third view, which can be expressed using what I will call the thesis of factual content, familiar from the previous chapter.

Factual content (F): Visual perception represents factual properties, which are properties that are in principle perceivable from multiple perspectives.

Factual properties include properties such as the shape, size, and kind of object. (F) is widely accepted today (Block 2003, Byrne 2001, Dretske 1995, Harman 1990, Metzinger 2003, Noë 2004, Schellenberg 2007, Searle 1983, Tye 2000).[2] Recall the distinction between factual properties and perspectival properties from above (section 2.1). (F) is a commitment to the idea that we represent factual properties and is silent on the issue of whether we represent perspectival properties.

As I hope to show, (PC) and (F) together make a strong case for (SA). In particular, (F) provides constraints for the consequent of (PC), thus making (SA) true. According to (PC), subjects anticipate some change. If (SA) is false, subjects anticipate no change in particular. Thus, (PC) and the denial of (SA) amount to the claim that any change due to self-generated movement would fulfill our anticipations. It seems, though, that there are a wide range of changes in visual experience that would not fulfill anticipations; there are many ways in which visual experience can change that would violate anticipations. Denying (SA) amounts to accepting that outlandish changes in visual experience can fulfill our anticipations.

Here is an argument for (SA) in form of a *reductio ad absurdum*. Assume that (F) and (PC) are true, but that (SA) is false. That is, the change anticipated from self-generated movement has no specificity whatsoever. Subjects only anticipate *some* change, but no particular change. Also assume, for the sake of argument, that it is possible to anticipate some change, but no change in particular. Imagine looking at a teacup on a table. The design on the cup looks interesting, so you move your head for a look at how the design continues on the hidden side of the cup. As an apparent result of this self-generated movement, the object on the table starts to appear as if it is a vase containing orchids, rather than a teacup. According to (PC) and the negation of (SA), you anticipated some change, but no particular change. Since the change from teacup to vase was indeed *some* change, (PC) and the negation of (SA) would suggest that seeing the vase fulfills your anticipation. Indeed, (PC) and the negation of (SA) seem to suggest that any experience that is not the same as the experience had just prior to the self-generated movement would fulfill your anticipation. The problem, though, is that there is no sense in which that change—from teacup to vase—fulfills any anticipation. Such a series of experiences would be *surprising*, and would not include the fulfillment of an anticipation. The possibility of surprise in perception offers a strong clue that anticipation has some specificity: as Dennett has noted, "Surprise is a wonderful, dependent variable" (2001: 982).[3]

The vase was not anticipated because seeing a vase is not compatible with the factual content of the experience prior to the self-generated movement, not compatible with the representation that there is a teacup (and not a vase) on a table. With a little imagination, one can see that there is no limit to the range of experiences that would be surprising, experiences that would not be anticipated when moving in for a closer look at the teacup. Since the negation of (SA) leads to the unwelcome conclusion that even the most bizarre visual experiences fulfill anticipations, we now have good reason to accept (SA).[4]

We have some specific anticipation because of (F). According to (F), we visually represent the world to be a particular way. As the example above should illustrate, we implicitly anticipate that the visual results of our self-generated movements will be coherent with the way we visually represent the world to be prior to those movements. Crucially, our visual anticipation need not include the *precise* way in which factual properties appear from different perspectives. As I discuss below, there is room for indeterminacy.

If my claims so far are correct, then (SA) looks to be correct also. According to (PC), we anticipate some change in visual phenomenology with self-generated movement. (F) gives good reason to think that this change will be of a particular nature, that this change will reveal new perspectives on the factual properties already being represented. If we accept both (PC) and (F), then competent perceivers should expect there to be a set of self-generated movements which will present different views on factual properties. More explicitly, (SA) can be shown to be true because now we have a more specific consequent for (PC). We have:

(PC') If S substantially changes her perspective on o, her visual phenomenology will present different views of o's factual properties.

With (PC'), the subject does not merely anticipate some change, she anticipates change that will reveal different views on factual properties. If I visually perceive an object to be a table from one perspective, I implicitly anticipate that it will continue to look like a table as I gain different perspectives on it. I would be surprised, and would have to reevaluate my representation of the environment, if the table ceased to look like a table as I moved to a different perspective on it.

The account that I am developing here shares some similarities with Charles Siewert's account of perceptual anticipation (2005). A crucial difference, though, can be found in my central claim, in (PC'). Siewert appears to suggest that the specificity of the sensorimotor conditionals can only be obtained through actual movement. He writes:

> It seems to me that I have no way of identifying how it looks to me as if it *will* look, *if* I do certain things, and what those 'certain things' are, other than the following. I actually *do* what I am disposed to do to get a better look at something, and take doing *that* as an illustration of the kind of activity that belongs in the antecedent of the conditional. (2005: 286)

Siewert is correct that actually taking a better look at an object is one way, indeed the best way, of determining how that object will look from another perspective. But Siewert appears to be making the stronger claim that actually taking a look is the only way to have a sense of how an object will look from other perspectives. I think there are reasons to resist this stronger claim. As Siewert acknowledges, it is possible that we can be surprised by the new appearances. Such surprises imply we have some anticipation of how things should appear when we move that does not depend on actually performing those movements.

The second reason to resist the stronger claim can be found in (PC') itself. Crucially, (PC') was derived based on the widespread commitment to (F). It is in (F), in the idea that we represent factual properties in perception, that one can find an alternative to Siewert's strong claim. One main way in which we can anticipate how things will appear without actually performing movements is because we represent factual properties. If I represent an object to have a particular factual property, then I anticipate that such an object will continue to appear to have that property as I gain different perspectives on it. I need not actually move in order to anticipate that new appearances of an object ought to be compatible with previous appearances of that object.

Note that there are probably perceptual experiences for which (PC') does not apply. Importantly, (PC') depends on the representation of factual properties, of properties that are in principle perceivable from multiple perspectives. Thus, in perceptual experiences with no factual content, (PC') would not apply. Many philosophers accept that normal experiences have some factual content. Can one have an experience with phenomenal character but no factual content? Perhaps afterimages and abstract geometrical hallucinations would be experiences of this sort (Metzinger 2003: 243).[5] If so, then these experiences would offer instances in which (PC′) does not apply. Similarly, the kinds of two-dimensional "cartoon-like" hallucinations commonly reported with Charles-Bonnet syndrome would be cases in which (PC′) does not apply (Ramachandran and Blakeslee 1998, discussed in Metzinger 2003: 238–239). When there is no perspectival connectedness, as in a hallucinated cartoon image, there will also be no anticipation of how those images will look from different perspectives. Of course, there still may anticipations of some kind in these cases, such as anticipations that afterimages will eventually fade. But these anticipations are different because they are not anticipations about how factual properties appear from different perspectives.

It may also be instructive to consider whether (PC) can ever fail. That is, are there situations in which subjects actively change perspective on an object, but would be surprised to find their phenomenology changing as a result? Such situations may occur in cases of mental illness such as schizophrenia. One symptom of schizophrenia is a loss of the sense of agency. Subjects self-generate movement, but they do not have the subjective experience as if this movement is self-generated. In such situations, the change in phenomenology can be surprising, because the subject does not have the sense of initiating the movement (Frith 1992, Gallagher 2005: Chapter 8).

Before turning to the objection borrowed from Siegel, I would like to make five points about visual anticipation, summarized in table 3.1, below. The first is that anticipation has degrees of determinacy. The second point is that anticipation is not limited to the hidden sides of objects. The third point is that perceptual anticipation need not be a deliberate mental act; rather, it is an ongoing feature of perception. The fourth point is that there are cases of visual experience in which the antecedent of (PC′) does not obtain. The fifth point is that visual anticipations are "stirred up" as we change our location and shift our attention. Below I elaborate on each point, with special attention to the fifth. These points especially are inspired by Husserl's work on perception (Madary 2012a; appendix, this volume).

Table 3.1
Five features of visual anticipations

1. Perceptual anticipation has degrees of determinacy.
2. Indeterminate visual anticipation is not restricted to the hidden sides of objects.
3. Visual anticipation is ongoing and not episodic.
4. The antecedent of (PC′) need not obtain.
5. Visual anticipations are continuously "stirred up" in the process of perception.

1. *Perceptual anticipation has degrees of determinacy.*

One can anticipate that the table will continue to look like a table from another perspective without anticipating the details of the granularity of the wood from that perspective. In such a case, the subject's anticipation is indeterminate with regard to the grain pattern, but the subject can reduce the indeterminacy by moving to take a close look at the grain pattern on the facing side of the table; anticipations have, as Husserl put it, "a determinable indeterminacy" (Husserl 1966 §1, also see Siewert 2005: 287).

2. *Indeterminate visual anticipation is not restricted to the hidden sides of objects.*

Notice that (PC) and (PC′) make no mention of hidden sides; rather, they are concerned with substantial changes in perspective on an object. There can be indeterminate anticipation for that which lies in the periphery as well as for details of the surface in the center of vision (Husserl 1966 §1). For observing peripheral indeterminacy, we have already encountered Dennett's card trick in the previous chapter (section 2.3). The indeterminacy within central vision simply refers to the fact that we can take a closer look at things (within limit, of course). To revisit the example from above, one can stand 4 meters' distance from a table and try to focus on a small area of the wood in the facing surface. From that distance, the grain of the wood may appear indeterminately, but one has the option of moving in for a closer look. We anticipate, usually implicitly, that moving closer will reveal details about the surface that were not available from 4 meters away.

3. *Visual anticipation is ongoing, and not episodic.*

We are capable of deliberately anticipating the appearance of something, but this capability is not what I am trying to describe here. The kind of visual anticipation that I mean here is a constant and ongoing feature of visual perception. We do not have to try to do it, and we cannot stop ourselves from doing it. When I claim that "I visually anticipate something," I do not mean that I make a deliberate decision to anticipate as I would, say, make a deliberate decision to dine at one restaurant rather than another. A related way to make this point is to say that visual anticipation is implicit. It is not something to which we usually attend, as we would attend to a decision about where to dine. Husserl uses the example of a patterned rug partially covered by some furniture (1900/1993: VI §10). We implicitly anticipate that the pattern of the rug continues under the furniture, but we might not do so deliberately. This feature of visual anticipation may strike some readers as being at odds with some accepted ways of thinking about mental states. I will address this kind of worry toward the end of the chapter, in section 3.4.

4. *The antecedent of (PC') need not obtain.*

Typically, the antecedent of (PC') will obtain. The truth of the antecedent follows from the fact that human vision is typically a process of active exploration. But there are some viewing conditions in which (PC') is merely counterfactual, conditions in which the antecedent is not realized. There are at least two kinds of cases here. First, the antecedent does not obtain when the subject does not substantially change her perspective on an object in view. This unchanging perspective can occur in two ways. First, the subject could cease to explore the environment with eye and body movements. We can stop visually exploring deliberately if we do not count microsaccades as substantial changes in perspective. Also, we can stop visually exploring as a result of a drug-induced paralysis (Aizawa 2007). Second, the subject could actively track a moving object, thereby keeping her perspective on the object the same. The second way in which the antecedent of (PC') would not be realized would be cases when visual exploration offers an initial view of an object.

How should we characterize these cases? Consider the first kind of case, in which the subject maintains fixation on the same perspective of an object. In such cases, (PC') would still hold, if only counterfactually. It is also reasonable to think that there are actual visual anticipations in these cases. Subjects in such cases would anticipate that appearances will not change if they maintain the same perspective on an object.

In the second type of case, in which we gain an initial view on an object, there is still plausibly some kind of anticipation, albeit perhaps a less determinate one. Siegel's (PC) was formulated with regard to an object, but we could interpret "object" broadly to mean an entire visual scene. That is, we can change the antecedent of (PC) from "If S substantially changes her perspective on o ... " to "If S substantially changes her perspective on the visual scene ... " without changing Siegel's argument. In the case in which we gain an initial view on something, the anticipation is probably best characterized as being something like Siegel's (PC), in which we anticipate some change, but no particular change. On the other hand, as I explain in the following section in more detail, the determinacy of visual anticipation is partly influenced by one's degree of familiarity with the visual environment. It is quite reasonable that a perceiver in a familiar environment can anticipate seeing particular objects at particular locations even when those objects are not currently in view. This issue is closely related to the question of which factual properties of an object are anticipated, a question that I address in my discussion of the next point.

5. *Visual anticipations are continuously "stirred up" (from Husserl's* erregen*) in the process of perception.*

There is a wide range of appearances objects can have, but these appearances are not all anticipated at once. There are at least two ways in which appearances of objects are excluded from visual anticipation. First, it is possible that appearances of factual properties currently in view are not anticipated. Second, it is possible that appearances of factual properties are not anticipated because those factual properties themselves are currently not visible. I will discuss these in turn.

First, there can be cases in which appearances of factual properties currently in view are not visually anticipated. For instance, as I walk down the street on a sunny day and look at a particularly attractive house, I anticipate how the house's appearance will change as I continuously move. I perceive the factual properties of the shape and color of the outside of the house. But I do not anticipate that the shape and color of the house will appear to me in ways far removed from my current perceptual situation. For instance, I would not anticipate that the shape and color of the house will appear to me as they would to someone who has leaped from an airplane and sees the house while free-falling, or how the shape and color of the house would look at night viewed by someone standing on her head. My perceptual situation and my self-generated movements constrain the anticipations of factual properties currently in view. The second way in which anticipations are constrained can be cases in which particular factual properties of objects are not visible. This second way can be illustrated by addressing a worry raised by Siegel.

Siegel has raised the worry that there may be cases in which it does not seem correct to include visual anticipations of hidden factual properties. She describes someone looking at a flowerpot but having the bizarre false belief that there is a miniature city on the hidden side of the pot (2010a: 180). Say the person with this strange belief has a visual experience of the pot, but that this experience does not include the part of the pot believed to contain the city. There is a sense, we might say, in which the experience of the flowerpot, or at least of the facing surface of the flowerpot, is veridical. After all, the subject with the strange belief does not hallucinate, or see any kind of illusion. On the other hand, (PC′) would seem to imply that the false anticipation of the miniature city is a part of perceptual content, which would bring some false element to the experience.

In its general form, this example brings out the problem of the *scope* of visual anticipation. There are a great number of factual properties in the world, and there is great variation in the amount of self-generated movement required to access those factual properties. Which factual properties are anticipated during normal visual experience? (PC′) suggests that subjects anticipate different views on o's factual properties. But this claim must be qualified. Objects have plenty of factual properties, and it would be wrong to claim that we anticipate all of them at once. For example, when a tall building catches my visual attention from the street, I can only vaguely imagine the details of the interior of the building. I have no visual anticipation about whether the building includes conference rooms, or about the interior design of any conference rooms. The interior would involve factual properties of the building, but these factual properties go beyond my visual experience.

On one hand, (F) and (PC) together suggest that we anticipate seeing different perspectives on some of o's factual properties. On the other hand, simple reflection shows that we do not anticipate *all* of o's factual properties. I am not sure that a precise line can be drawn between the perspectives on properties that are anticipated and those not anticipated, but I think we can arrive at a rough description. I have already noted above that not all appearances of the factual properties currently in view are anticipated at once. My support for this claim appealed to the way in which my current viewing conditions and self-generated movements would constrain the way in which a house would appear to me. Now, with Siegel's flowerpot example, we have a slightly different issue. The question now concerns which factual properties are anticipated. A thorough answer to the question of which factual properties are anticipated and which are not may require further investigation, but here is a brief account.

Which factual properties are anticipated? Roughly, one's current perspective determines the scope of visual anticipation: If the subject sees a factual property of the object from one perspective, she will anticipate self-generated movement that gives new perspectives on the object will reveal new perspectives on *that* factual property. If I see the table from one perspective, I will visually anticipate that it will continue to look like a table as I move around it. This claim does not entail strange commitments in the examples from above; since neither Siegel's miniature city nor the conference rooms of the tall building are factual properties that were in the subject's current perspective, we need not claim that perspectives on those properties were anticipated. Siegel's worry can be addressed by claiming that false beliefs about hidden sides do not entail visual anticipations of the contents of those beliefs.

But the problem of describing which anticipations are stirred up is not that easy to solve. The claim in the preceding paragraph is incomplete because, as discussed in the previous point, there are times in which we anticipate that self-generated movement will reveal an *initial* perspective on an object or a factual property. If Siegel's subject with the strange belief about the city decides to move in order to view the city, then there is a point at which she will anticipate seeing the city, and, according to the example, her anticipation will be disappointed (since there is in fact no city there). As she moves closer to the place in which she thinks she will see the city, her anticipations of the city will be stirred up. To the extent that those false anticipations are stirred up, there is a weak sense in which her experience is nonveridical. Perhaps more interestingly, there will also be some kind of perceptual conflict at the instant in which she thinks the city should be visible to her. At this instant, the anticipation of the city is in conflict with the sight of a smooth, cityless flowerpot. I am not sure whether this instant of conflict should count as nonveridical, but this uncertainty bears little on (PC').

Another complication is that there are common viewing conditions in which multiple objects are simultaneously in view. In order to accommodate such situations, I should add that (PC') need not be limited to one object at a time. That is, changing one's perspective according to one's interests, goals, and exogenous visual attention might bring different views on the factual properties of multiple objects. Due to the nature of our visual field, in which only the center of vision is determinate, increasing the number of objects in view increases the visual indeterminacy. Along the same line, there might be cases in which it is not clear how to individuate visual objects. For instance, recall the example of the tall building from above. One might say that the building and its interior is all one object. But one might also want to individuate objects more finely, and claim that each room in the building is a visual object that in turn contains more visual objects. The arguments for (PC) and (PC') do not depend, from what I can tell, on a precise way of individuating visual objects.

These considerations lead to the question of the relationship between perceptual anticipations and beliefs. For example, one might be told, before entering a familiar room, that there is a surprise hidden inside the room. One might suspect a gift or an unexpected visitor. This kind of case suggests that there may be mental states that are not easily categorized as either visual anticipations or beliefs. One visually anticipates seeing something unusual upon entering the room, but this anticipation is far more indeterminate than the anticipation, say, of how one's bike will look as one walks around it. The anticipation about the surprise in the room is like a belief in the sense that it is not, or is only minimally, constrained by one's current perceptual context.[6]

So far in this chapter I have defended the thesis of specific anticipation (SA), and combined (SA) with Siegel's (PC) to introduce (PC'). I have also introduced five features of visual anticipation and have developed the fifth feature in order to address a possible worry inspired by an example from Siegel. Now I turn to some consequences of (PC').

3.3 Variation in Perceptual Content

If (PC′) is true, then there can be variation in perceptual content under identical viewing conditions. Here is a vignette to illustrate this claim for variation in content for two different subjects.

One afternoon at a party, Lily, unimpressed with her company, took a fleeting interest in a sculpture in the corner of the living room. A few minutes later, coincidentally, Rosemary took a look at the same piece of art. Even more of a coincidence, Rosemary viewed the sculpture from the exact same place in which Lily viewed the sculpture. In an astonishing coincidence, Lily and Rosemary viewed the sculpture with the same pattern of saccades for the same length of time. For the periods in which they each viewed this sculpture, did Lily and Rosemary's visual experiences have the same content?

Here are some details relevant for answering this question. Both Lily and Rosemary have normal vision, so there was no variation due to a general difference in the visual system. Also, neither of them had ingested any substances that might have altered their visual experience (it wasn't that kind of party). The two women's eyes were at the same height that day (Lily is slightly shorter than Rosemary, but was wearing higher heels), so fine-grained differences in perspective would not have made a difference either. Similarly, their bodies were positioned in the same manner. They both attended to the sculpture and focused their eyes on the surface of the sculpture. Nothing in the visual scene changed between viewings; no one pulled the shades or vandalized the sculpture.

With these details in place, I take it that many would answer my question in the affirmative: yes, surely Lily and Rosemary had the same content of their visual experiences. Why? Mainly because there are intuitions, likely widespread, to the effect that all of us perceive the same world in pretty much the same way. I take the following passage from David Lewis to illustrate this thought:

> The scene before my eyes causes a certain sort of visual experience in me, thanks to a causal process involving light, the retina, the optic nerve, and the brain. The visual experience so caused more or less matches the scene before my eyes. All this goes on in much the same way in my case as in the case of other people who see. (1980: 239)

More recently, Susanna Schellenberg has expressed roughly the same idea with regard to seeing a cup:

Anticipation and Fulfillment

Any perceiver occupying the same location would, ceteris paribus, be presented with the cup in the very same way. (2008: 61)

I am not suggesting that these passages commit Lewis or Schellenberg to any particular thesis about variation in perceptual content. What I am suggesting is that these passages express the reasonable thought that we all pretty much see the world in the same manner.

If (PC') is correct, though, there can be variation in perceptual content under these identical viewing conditions. Consider further details about Lily and Rosemary. As it turns out, the sculpture's creator, Jack, is a close friend of Rosemary's. She is familiar with his work, and owns a few of his pieces. Two of the pieces she owns are similar to the piece in the corner of the living room at the party. Lily, on the other hand, had no idea who sculpted the piece in the corner, had little knowledge of or interest in contemporary sculpture, and had never before seen anything much like the artwork in the corner.

As a competent perceiver, Lily has visual anticipations about the sculpture. She anticipates that it will continue to look like a novel sculpture as she takes different perspectives on it. But those anticipations are indeterminate, much more indeterminate than Rosemary's anticipations, since Rosemary is familiar with Jack's artistic style. For instance, Rosemary is familiar with a new technique that Jack has started using on the surfaces of his pieces and is able to anticipate the peculiar way in which the specular highlights will change as she moves. One apparent consequence of (PC'), then, is that different subjects can, and often will, have different perceptual content under identical viewing conditions. Perceptual content is rich in the sense that it includes a unique contribution from each perceiver.

Variation in perceptual content is not limited to different subjects. If (PC') is true, then there can also be variation in perceptual content within the same subject under identical viewing conditions.[7] Upon leaving the party that afternoon, Lily shared a cab home with one of the party goers. It turned out to be Jack, the artist. Lily and Jack entered into a serious romantic relationship, which naturally involved her becoming more familiar with his work. A year later, Lily and Jack were back at the house for another party and Lily viewed that same sculpture in the corner, with the same pattern of saccades. But this time, of course, she viewed it with much more determinate visual anticipations; she now knew Jack's work well. Both the viewing conditions and the perceiver were the same, but the perceptual content was different.

In these two cases of variation of perceptual content, the variations in content were due to the subject's level of *familiarity* with the object. The idea that familiarity with a scene can influence perceptual experience has been explored both by Siewert (1998: 257–258) and Siegel (2006: 493). What I offer here is an account of *why* familiarity changes experiential character: when we are more familiar with a scene or an object, we have more determinate visual anticipations of how things will look as we move.[8]

Here I have given cases of variation under the unrealistic constraint that the self-generated movements with which Lily and Rosemary visually explored the sculpture were the same. If we loosen this unrealistic constraint, we can see there are factors besides familiarity that produce a variation in content. For instance, it is well known from classic eye-tracking research that a subject's task influences saccade patterns (Yarbus 1967; section 5.3 below). Also, it is reasonable that a subject's attentional or emotional state can influence the precise way in which she explores a visual scene, as well as influence the determinacy of her visual anticipations.

3.4 Visual Anticipation and Two Distinctions

In this chapter I have used examples in order to familiarize the reader with what I am trying to describe as visual anticipation. But there are still some issues that I have not addressed. For instance, one might ask where visual anticipations fit within the distinction between personal and subpersonal mental states. One might also ask where visual anticipations fit within the distinction between perception and cognition.

Many philosophers have made use of a distinction between personal and subpersonal levels of explanation (following Dennett 1969). The former includes mental states that we commonly attribute to persons, such as intentions and beliefs. The latter includes the causal mechanisms that enable various types of intelligent behavior. On one hand, visual anticipation seems to occur on the personal level: my arguments for such anticipation depend on first-person reflection, rather than empirical modeling. On the other hand, it is not quite correct to claim that visual anticipation is something we *do* deliberately. As I claimed above, it is ongoing, and not episodic. What may be most relevant here is that if we take the subpersonal

Anticipation and Fulfillment 57

description to mean some kind of causal description, then visual anticipation, as described above, is not subpersonal. Perhaps the best understanding of visual anticipation is as a nonstandard instance of personal-level description. The truth of (PC′) may put pressure on the distinction itself. In any case, I am happy to concede that more clarification is needed here, and I revisit this issue briefly in chapter 8 to that end.

What about the distinction between perception and cognition? Is visual anticipation a part of vision, or is it better understood as a part of cognition? Are anticipations a particular kind of belief? Here it is important to keep in mind that the distinction between perception and cognition has been used because some have found it helpful for categorizing mental states and processes (also recall the discussion in section 1.1 above). The usefulness of the distinction in other domains does little to speak for or against the reality of visual anticipation or the truth of (PC′). Perhaps the truth of (PC′) indicates that there are mental states that are somewhere in between paradigmatic perceptual and cognitive states, such as one's belief that there is a surprise hidden in a room (discussed above). One may want to use a liberal notion of "belief" in order to have visual anticipations count as beliefs. I do not see that much turns on this issue. The important point is that (PC′) stands or falls based on independent arguments, not on the fact that it might not fit clearly with an existing distinction. In the final chapter, I will make the case that a good bit of what has traditionally been relegated to the domain of cognition can be better understood as being embedded within the cycle of action and perception. If this last claim is true, then the traditional distinction between perception and cognition should be reconsidered.

3.5 Summary

In this chapter I have provided an argument for (PC′). Visual phenomenology includes anticipation with various degrees of determinacy. One consequence of (PC′) is that there can be variation in visual content for identical viewing conditions.

To return to the larger picture, recall premise (1) of my Main Argument:

(1) *The descriptive premise*: The phenomenology of vision is best described as an ongoing process of anticipation and fulfillment.

With the previous and the present chapters, I have made my case for premise (1). First, the structure of anticipation and fulfillment accommodates the three constraints presented in the previous chapter. Second, (PC′) establishes that visual phenomenology is anticipatory.

Before moving on to my case for the empirical premise, for premise (2) of the Main Argument, I would like to further illustrate premise (1) by relating it to the way in which visual content is often understood in contemporary philosophy of perception. The following chapter covers this theme.

4 The Question of Content

Following the convention in the philosophy of mind, if a state has *content*, then it has information with accuracy conditions; it represents things to be one way or another. This representation can be either accurate or inaccurate (Peacocke 1983, Fodor 1987, Dretske 1995). One of the questions in the philosophy of perception, then, is whether perceptual experience has content. Does experience—visual experience in particular—represent the world to be one way or another? And if so, what is the nature of this content? Philosophers have defended a number of different answers to this question in recent years, too many to cover here in any depth.[1] My premise (1), that the phenomenology of vision is best described as an ongoing process of anticipation and fulfillment, suggests a specific answer to the question of visual content. It suggests that visual experience does have content, but that this content is, in some ways, unlike other kinds of content found in the literature.

This chapter has three aims. First, I will further illustrate the nature of visual content which follows from (1); let us call it *AF content*. Then, I will contrast AF content with other descriptions of visual content in the literature. Finally, I will show how AF content helps to solve three outstanding puzzles in the philosophy of perception. Unfortunately, AF content does face a well-known epistemological problem. I suggest ways in which this problem can be addressed as well.

4.1 Introducing AF Content

In order to introduce AF content, first recall the argument from the previous chapter. There I argued, using a *reductio*, that Siegel's (PC) together with (F) leads to (SA) and (PC'). Recall that (F) is the widely accepted thesis of factual content; it is the claim that visual perception represents factual properties, which are properties that are in principle perceivable from multiple perspectives. Thus, my argument for (PC′) already contains a premise to the effect that there is visual content of a particular kind. But there is more to be said, for I have not yet specified the way in which factual properties are represented in perception.

If (PC′) is correct, then we visually anticipate the sensory consequences of self-generated movements. When things are going well, these anticipations are fulfilled to various degrees. Since visual anticipations can be correct or incorrect, they have content. *AF content, then, is the content of the visual anticipations themselves.* AF content is determined both by unfulfilled and fulfilled anticipations. Due to the fact that our eyes can only focus on a fraction of the visual scene at each saccade, much of our visual content is made of unfulfilled or empty anticipations. Fulfilled anticipations are always fleeting as new anticipations are continuously stirred up.

Although my emphasis has been on the future-directedness of visual perception, I would also like to suggest that fulfilled anticipations are retained as they fade into the past (as in Husserl 1966 §10–12). That is, fulfilled anticipations linger in consciousness as they partly determine the content of one's future anticipations. In order to perceive objects as unified, there must be a synthesis between the "just past" fulfilled anticipation, which becomes a retention, and the upcoming fulfillment of a new perspective on the object. There is surely much more to say on this synthesis of fulfillments, but doing so would require a lengthy detour into Husserlian phenomenology (see Husserl 1900/1993 VI, for example).

As discussed in the previous chapter, the content of anticipations depends, in a fine-grained way, on two factors. First, it depends on which factual properties the subject has in view. Second, it depends on the background and context of the perceiver. Consider a visual experience of a vase. If I see a vase for the first time that appears to be uniformly green from my perspective, but it really has a vertical red stripe on the occluded side, then my anticipation of seeing green on the hidden side of the vase is incorrect. As I move to look at the hidden side, I anticipate seeing green and this anticipation is disappointed when I begin to see the red stripe. The anticipation of seeing green at exactly the area when I begin to see red is incorrect. The example of the vase is meant to illustrate, in a basic manner, the way in which (PC′) leads to AF content. With the basic idea in place, I will, in the rest of this section, make two more points about AF content and address the way in which AF content can deal with two special cases.

The first point has to do with the relationship between AF content and (F). At first blush, there seems to be a tension between the two, since (F) has to do with factual properties, and AF content has to do with the way properties appear from particular perspectives. Here it is crucial to keep in mind that the thesis of anticipation and fulfillment, or AF from the Main Argument, is formulated precisely as an attempt to accommodate the fact that we represent factual properties, but only from particular perspectives. There is no tension between AF content and (F) if we qualify (F) with the claim that visual representation of factual properties is always *incomplete*. That is, we do not anticipate seeing objects from all perspectives and in all possible lighting conditions all at once. As discussed in the previous chapter, which anticipations are stirred up depends on viewing conditions as well as psychological features of the perceiver. We represent factual properties through visual anticipation of the way those properties will appear within the context of a particular perceptual episode.

Here I should add another comment about the relationship between AF content and (F). One might be tempted to interpret AF content as a return to phenomenalism of the mid-twentieth century (C. I. Lewis 1946). To be clear, I am not proposing that factual properties are reducible to the various ways in which they appear to us. Nor am I making the weaker phenomenalist claim that the perception of factual properties is reducible to a set of anticipations and the fulfillments of those anticipations. Visual anticipations depend on there being a representation of a factual property, not the other way around. My anticipation that the sphere will continue to appear red as I shift my perspective on it depends on my intentional state, which is directed toward the color of the sphere qua factual property. As mentioned in the previous chapter, if I cease to visually experience factual properties, then I cease to visually anticipate in the sense expressed using (PC'). Hallucinations of abstract geometrical patterns provide a good example of this possibility, as mentioned in the previous chapter.

The second point I would like to make is that the anticipations that underlie AF content are always fused together; they are *verschmolzen*, to use Husserl's term (see 1900/1993, III §8, for example). This claim means that visual anticipations cannot be individuated, they flow into one another without distinction. The continuous nature of visual phenomenology requires that this be the case. Because of this fusion of visual anticipations, AF content will always escape a complete verbal characterization. Along the same lines, recall the discussion of the indeterminacy of visual anticipation in the previous chapter. If visual anticipations have various degrees of determinacy, then we should not expect natural language to capture fully the content of anticipations. If visual anticipations are both fused together and indeterminate, then AF content is, in a word, messy. Some might find the messiness of visual content to be undesirable, but the truth is not always as we'd like it to be. More importantly, as I argue below, treating visual content along the lines I am suggesting solves outstanding puzzles in the philosophy of perception. Denying the messiness of visual content creates needless puzzles.

Now I shall further illustrate the nature of AF content by treating two special cases of visual experience. The first case has to do with cases of illusion and hallucination. My general strategy is the same for both illusions and hallucinations, but let us consider illusions first. At first it may not be clear how my general framework might apply to cases of illusion. When we look at common visual illusions—see the Müller-Lyer illusion in figure 4.1 below, for instance—we seem to experience the illusion from a single perspective. Unfulfilled visual anticipations about different perspectives seem irrelevant as an explanation for the illusory content.

The important point to make about visual illusions is that *fulfilled anticipations are not sufficient for veridicality.* When you first saccade onto the illusion, you might have anticipations of seeing some figure. These anticipations are fulfilled. Many readers will be familiar with the illusion and have determinate anticipations about two horizontal lines with tails. Some

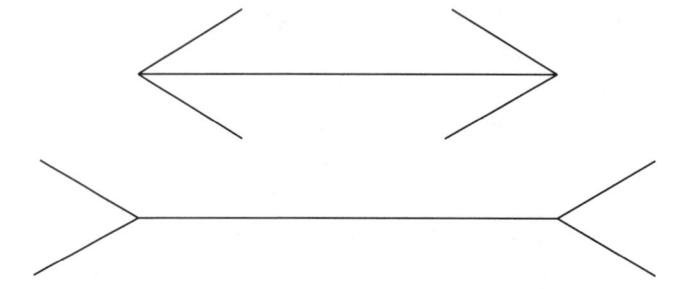

Figure 4.1
The Müller-Lyer illusion.

readers may even have had implicit anticipations of seeing the illusion in this book before it was even mentioned in the text—philosophy books on vision often include it. Now, seeing figure 4.1 generates the implicit anticipations that you will continue to see figure 4.1 if you look at the page (or the screen) from different perspectives, perhaps from a distance of two meters, or from an unusual angle. All of those anticipations are accurate and can be fulfilled if you so choose. If you visually judge the lines to be of different lengths, as most of us will, then you also form other anticipations. You may anticipate that your saccade from the right end point of the top line to the right end point of the bottom line will trace a path that is not at a right angle relative to the horizontal lines. This anticipation is false, since the end point of the bottom line is directly below the end point of the top line. Even though this anticipation is false, it can be fulfilled since it seems to you, so long as you are a victim to the illusion, that the tip of the top line is not directly above the tip of bottom line. It is for this reason that I claim that fulfilled anticipations are not sufficient for veridicality. Importantly, your visual judgment that the lines are different lengths also generates anticipations about how the lines will appear when a ruler or some measuring tool is held against them. (Such anticipations may not be stirred up unless you take steps toward taking the measurement.) These anticipations are not accurate, and these anticipations are relevant for explaining the illusion. They will not be fulfilled if you take the measurement, and it is for precisely this reason that you would conclude that figure 4.1 is illusory with respect to the relative length of the lines.

This analysis of AF content and visual illusions raises the interesting issue of the role of action in visual illusions. Since human vision is typically a process of active exploration, there should be connections between saccade patterns and the effect of visual illusions (as well as ambiguous figures). This issue has already been explored empirically to some extent (Festinger, White, and Allyn 1968, Troncoso et al. 2008), but I suspect that there remains a good bit of unexplored territory here (I return to this theme in section 7.4 below).

Having discussed how AF content applies to illusions, now consider how it can apply to hallucinations. Again, the important point is that fulfilled anticipations are not sufficient for veridicality. A subject might have a visual hallucination of a dagger from a particular perspective. If he does not perceive the dagger as hallucinatory—that is, if he does not realize that he is hallucinating—then the subject would form implicit anticipations that the dagger is perceivable from other perspectives. These anticipations might even be fulfilled. After the hallucinatory episode, the subject would then have other visual experiences, other sequences of anticipations and fulfillments that reveal the dagger was merely a hallucination. Perhaps the subject begins to see a flat, empty table where the dagger was hallucinated to have been. Closer inspection reveals no sign of a dagger anywhere in the vicinity. Perhaps the subject also regains some cognitive abilities that had been clouded during the hallucinatory episode. The important point is this: subsequent—in this case post-hallucinatory—anticipations and fulfillments can render false previous series of anticipations and fulfillments. The previous fulfillments alone were not sufficient for veridicality. Of course, it is also possible hallucinating subjects never have a subsequent series of anticipations and fulfillments that conflict with the hallucinated content. In such a case, the subject may never discover that he had hallucinated.

There is probably more to be said about AF content and illusions and hallucinations. But what I hope to have shown is the general form of the way one can treat these cases. There remain issues about how AF content can be applied to describe particular illusions—the Müller-Lyer is only one of many. Similarly, there remain fascinating issues about different kinds of visual hallucination. There are conceptual distinctions about the different kinds of hallucination that are experienced by humans (Shannon 2003), and there is also the philosopher's thought experiment about pure or perfect hallucinations, which are supposed to be subjectively indistinguishable from a veridical experience (Martin 2002). My general reply should be applicable to all of these cases.

Having dealt with cases of illusion and hallucination using AF content, I now turn to a final case for this section. In particular, I will consider how AF content might account for cases in which we do not actively explore the environment. As mentioned in the first chapter, Aizawa, in his critical assessment of Noë's work, has referred to patients in a drug-induced paralysis who report having visual experiences (Aizawa 2007, Schwender et al. 1998, Sandin et al. 2000). These patients are not actively exploring, yet they claim to have visual experiences. AF content is based on (PC′), which has to do with the consequences of self-generated movements. These cases might at first appear to challenge AF content, for the following reason: since there are purported cases of visual content without self-generated movement, there must be something more to visual content than AF content.

This concern can be dismissed by recalling the discussion in the previous chapter (see point 4 of section 3.2), about (PC′) when the antecedent does not obtain. In this case, the antecedent does not obtain due to paralysis. Importantly, (PC′) can still determine AF content counterfactually. Thus (PC′) is completely compatible with paralyzed patients having visual content. Of course, they are not able on their own to change their perspective on o, but it should still seem to them that, if they could, they would have different views on o's factual properties.

Now I have presented a basic account of AF content, I turn to consider other theories of content in the literature. I suggest that they all suffer shortcomings where AF content does not.

4.2 Alternative Theories of Content and Their Shortcomings

Thinking about vision as having AF content has several advantages over other, more commonly found, ways of thinking about visual content. In this section, I will present some of the alternatives and describe their shortcomings. AF content has the advantage of being motivated by the structure of perception itself, by the three constraints from chapter 2. The more common ways of thinking about visual content do not share this advantage. Also, as I show in the next section, AF content is able to avoid a number of puzzles in the philosophy of perception, where these more common approaches cannot. The most common way of thinking about perceptual content in contemporary philosophy is to think of it as propositional content. First I will discuss the propositional account of visual content, and then turn to object, picture, and scenario content.

Many philosophers think that the content of visual perception is best expressed using the following form:

S sees that P.

I will refer to the view expressed by this claim as *perceptual propositionalism*. Perceptual propositionalism is often taken to mean that the content of perception is much like the content of a belief or judgment, content that one might express using one's natural language. For example, I judge that the robin is on the tree branch because my visual experience represents that the robin is on the tree branch; or, to put it more simply: I see that the robin is on the tree branch. Perceptual propositionalism could also mean that the propositional content of perception differs from the propositions that one would express using natural language, perhaps due to the level of complexity. One's particular type of perceptual propositionalism will depend on one's preferred theory of propositions, on whether one thinks that propositions are Fregean, Russellian, sets of possible worlds, and so on (Siegel 2015, section 3).

One of the most well-known claims of support for perceptual propositionalism comes from John Searle:

The content of the visual experience, like the content of the belief, is always equivalent to a whole proposition ... all seeing is seeing *that*: whenever it is true to say that *x* sees *y* it must be true that *x* sees that such and such is the case. (1983: 40)

Searle's propositional account of perceptual content grows out of his analysis of belief as a prime example of an intentional state. He points out that his propositionalism about perceptual content "does not imply that visual experience is itself verbal" (ibid.). But his position does suggest that the content of visual experience is that of a whole proposition, though the proposition may not be realized linguistically. Support in favor of a propositional account of perceptual content can also be found in the work of Armstrong (1968: chapter 10), Peacocke (1983: chapter 1), McDowell (1994: chapter 2), Williamson (2000: chapter 9), and Byrne (2001). Perceptual propositionalism is currently orthodox in the philosophy of perception.

Despite its widespread acceptance, I should mention that one important voice of dissent from perceptual propositionalism comes from Tyler Burge. Burge mounts a sustained attack on some of the philosophical underpinnings of perceptual propositionalism from the history of analytic philosophy, taking on figures such as Strawson, Evans, Quine, and Davidson (2010: chapters 6 and 7). I will not review his arguments here, partly because they are directed at specific views espoused by the philosophers just listed, and partly because there are elements of Burge's larger project that are in tension with my own, notably his skepticism about first-person description for understanding perception (2010: 131–133). But I do note that his anti-propositionalist arguments may be, at least to some extent, complementary to my view. One such critical point from Burge is that a wide variety of animals have perceptual systems, but very few animals are capable of having propositional attitudes (2010: 538).

Apart from Burge's attack on perceptual propositionalism, let us consider three general reasons to be in favor of perceptual propositionalism from the standpoint of contemporary philosophy of perception. Importantly, these reasons are not based on the nature of perceptual experience, not sensitive to the three constraints introduced in chapter 2. The first reason is based on a presupposition about mental states in general; the second is based on the correctness conditions of perceptual content; and the third is motivated by an epistemic worry.

One might be attracted to perceptual propositionalism if one adopts a methodology that approaches all mental states in terms of some kind of linguistic structure. For instance, one might think that all mental states are propositional attitudes (a view targeted for criticism in Montague 2007). Similarly, one might accept Michael Dummett's axiom of analytic philosophy, that "the only route to the analysis of thought goes through the analysis of language" (1996: 128). Of course, thought is not the same as perception, but both are faculties of mind. Or, finally, one might point out that we typically report perceptual content using propositions. These kinds of reasons are not really reasons at all, but are methodological presuppositions that involve, as Dennett puts it, "misprojecting the categories of language back onto the activities of the brain too enthusiastically" (1991: 365). Beginning with language in the study of the mind is a presupposition in a rich tradition, but it is not motivated by a consideration of perception itself, and is therefore not compelling. It is a method that, in Burge's words on Quine, fails to appreciate "that the semantics of language is initially determined by perception" (2010: 215). Philosophy of perception should begin with perception itself, not language.[2]

A second motivation for perceptual propositionalism can be formulated as follows. If perception has content, then it has truth conditions. In classical logic, the bearers of truth are propositions. Therefore, the content of perception is propositional. The problem with this reasoning is that content has accuracy conditions, not necessarily truth conditions (McGinn 1989, Crane 2008: 6–7, Rowlands 2010: 115). A map or a picture or a model can be more or less accurate, but these things are not themselves true or false. It is most natural to describe the content of a map or picture as imagistic, not propositional.

The third motivation for perceptual propositionalism is an epistemic worry found in the work of Donald Davidson (1983/2001) and John

McDowell (1994). The argument, roughly, is as follows. Our beliefs about the perceived world are propositional attitudes. The only thing that can rationally justify a propositional attitude is another propositional attitude. Thus, if beliefs are rationally justified by perceptual content, then perceptual content must be a propositional attitude. Call this argument the Davidson/McDowell worry; I will discuss it later on.

The main shortcoming of perceptual propositionalism is that it is not motivated by claims about perceptual experience itself.[3] Of the three reasons commonly given in favor of it, the first two can be safely ignored. They can be ignored, I suggest, because our understanding of perceptual content should be driven by the nature of perception—not by presuppositions influenced by twentieth-century philosophy of language. The third reason in favor of perceptual propositionalism, the Davidson/McDowell worry, does strike me as being a real concern. I will return to it below.

Now consider another way of thinking about visual content, what I will call the *object-relational* view. One powerful way to challenge the idea that all mental states are propositional attitudes is to give examples of mental states that are better understood as relations, rather than in terms of propositions. Love and hatred are good examples of such states. To use Tim Crane's example, "Napoleon's love of Josephine is not a propositional attitude" (2008: 12).

One alternative to the perceptual propositionalism is the claim that perception, like love and hate, is a relationship between a perceiver and an object (Vision 1996: chapter 4). We can call this alternative the *object-relational view*. The propositionalist would claim

S sees that P.

An advocate of the object-relational view would claim

S sees O.

Unlike propositionalism, for which accuracy conditions are clear, the object-relational account of vision does not have obvious accuracy conditions. For philosophers who reject the idea that perception has content, the lack of accuracy conditions may be a welcome advantage to the object-relational strategy (see the following section). Michelle Montague has suggested that an object-relational strategy can offer accuracy conditions.

For why shouldn't one say that the condition of satisfaction of a visual experience of an object X ... is just that one did in fact see X, and that its conditions of satisfaction consist simply of the object X plus a certain sort of causal-connection condition? (2007: 514)

This kind of approach has some similarities with what I will be developing, but it still has too much similarity with the methodology of the perceptual propositionalist. Both of these approaches look to constructions in natural language as the primary guide to perceptual content. This methodology obscures the actual structure of perceptual experience.

A third approach to perceptual content would be to treat it as analogous to a picture or an image. Mohan Matthen and Tim Crane follow this strategy to an extent. Matthen suggests that the content of visual perception can be "quite similar" to the content of pictorial vision (2010: 114), and Crane uses the analogy to argue against propositional accounts of perceptual content (2008).

Both Matthen and Crane are clear that the analogy breaks down at a point. Matthen, for instance, emphasizes that pictures do not offer information about the location of objects relative to the perceiver (2010: 115). Here I want to add another point of disanalogy between perception and pictures: in the case of looking at a picture, there is no sensorimotor connection between the subject and the objects in a picture. When visually perceiving objects in the world, there is always the possibility, in principle, of taking a different view on those objects. In pictures, there is no such possibility. Of course, there is the possibility of taking a different point of view on the picture itself, but there is no possibility of taking a different view on the objects in the picture. Thinking about perception in terms of pictures or images should be avoided because perception involves changing perspectives and is dynamic.

The fourth way to think about visual content is as *scenario content*, and has been developed by Christopher Peacocke. Scenario content is the way space is filled out relative to the fixed axis of a perceiver (1992: 63). With this suggestion, Peacocke goes some distance in accommodating the perspectival nature of perception, but he does not address peripheral indeterminacy, and he avoids the temporal nature of perception altogether with the unsatisfying assertion that "perceptual experience has a present-tense content" (1992: 65). AF content goes a step beyond scenario content by including peripheral and temporal indeterminacy, and by including temporality (following Almäng 2014, for instance) in the structure of perceptual experience itself.

4.3 On the Denial of Perceptual Content

So far in this chapter, I have covered a number of views that are all compatible with (F). AF content, perceptual propositionalism, and the other views just covered are all in agreement that visual perception represents factual properties. In this section I would like to discuss a family of views that all reject (F). The literature on this topic has grown vast in recent years and can be rather confusing, so I will be glossing over quite a few details. The denial of (F) in contemporary philosophy of perception goes back several decades (Hinton 1967, 1973, Snowdon 1980/1981, McDowell 1982) to a position known as disjunctivism. That position is so called because it maintains that experiences are *either* of the veridical kind *or* of the nonveridical kind, such as illusions and hallucinations. Now, recall that (F) is the claim that we *represent* factual (nonperspectival) properties in perception. Following the convention in philosophy of mind, a *representation* is a state that might be either accurate or inaccurate, that might feature in a veridical or a nonveridical experience (Peacocke 1983, Fodor 1987, Dretske 1995). The disjunctivist denies that perceptual experience includes any state that could be common to both veridical and nonveridical experiences. By denying that there is a representational state shared by veridical and nonveridical experiences, the disjunctivist rejects (F).

There are a number of different variations of disjunctivism (covered, for example, in Soteriou 2014), but the core idea is captured nicely in the following passage from Charles Travis:

It may be the pig's presence that makes me aware that there is a pig before me, thanks to my *seeing* it. I do not then *erroneously* take it to be there. If there is a pig to do any misleading, then that is one score on which I am not misled. Equally for my seeing one. If I am misled into taking a pig to be before me when there is none, I am misled by something else. The rear half of the pig, protruding from behind the barn, might do that if there is only a rear half there (perhaps mechanically animated). It *would* do so if I took it to mean there was a whole pig, thus inferring what, even if indicated, was not so. (If it is not just a rear half, perhaps one sees a *pig*, rear view.) Indications of a pig, or what I take for such, may lead me to conclusions. That is not for them, or anything, to *represent* something to me as so. (2013: 33)

To put things in a clearly disjunctive format, *either* one sees a pig *or* one is misled into thinking that one sees a pig. There is no visual representation as of a pig common to both the veridical and nonveridical experiences.

Although disjunctivism is a rejection of perceptual propositionalism, both share a common methodological presupposition: Both approach perception through natural language. Travis is explicit about the strong influence of J. L. Austin's ordinary language philosophy on his own view, and other disjunctivists are explicit about their desire to preserve "commonsense" reports about perception (Brewer 2011, for example). One consequence of this methodology is that the disjunctivist framework often tends to leave out the basic distinction, introduced in chapter 2, between the perspectival and the factual elements of perception. This distinction is not commonly found in everyday perceptual reports, but it is nonetheless a universal feature of visual perception. And it is this distinction that can help address the concerns that motivate disjunctivism without requiring the rejection of (F).

The relevant insight to take from the perspectival nature of perception is that there is a sense in which factual content always goes beyond what is currently in view. As Husserl put it, "external perception is a constant pretension to accomplish something that, by its very nature, it is not in a position to accomplish" (Husserl 1966 §1 / 2001: 39). To put it differently, we never see a pig, or anything else, *simpliciter*. We always see factual properties, such as being a pig, from a particular perspective, which means that the way the objects appear from other perspectives is necessarily excluded. Factual content, on the view I have been developing, always involves unfulfilled anticipations about what is beyond one's current perspective. Perceptual experiences are always, in this sense, incomplete.

This incomplete feature of visual experience serves as a response to Travis' main complaint about perceptual content, and thereby defuses a main motivation for his disjunctivism. Travis attacks perceptual content views that are committed to experiences reporting a determinate way things are, and experiences having a particular "face value" P that we can accept or reject (2013, chapter 1). In short, Travis attacks perceptual propositionalism.

If visual experience is always incomplete (and indeterminate) with regard to its factual content, then Travis' skepticism about there being "*the* way things appear to be" (2013: 26) poses no problem. Instead of *the* way things appear to be, AF content is indeterminate and messy, always made up of an ongoing process of anticipation and fulfillment. Those anticipations have content because they can be disappointed when, for example, we move in for a different perspective. There is no single way that visual experience presents things to be, but there is content to visual perception nonetheless.

Consider Travis' example of the pig once again. On his analysis, either the appearances do not mislead (when there is really a pig in view) or they do mislead (when, for example, there is a mechanically animated artificial pig's rear in view). The important point brought out by my discussion of AF content is that, even when appearances do not *mis*lead (when there really is a pig in view), those appearances still *lead* to, or stir up, unfulfilled anticipations about how the pig would look from other perspectives. Even in the veridical case of seeing the pig, there are still unfulfilled anticipations, and those anticipations have content. One possible reply for Travis here would be to deny that unfulfilled anticipations are an element of visual experience. I will return to this reply below, after discussing some concerns of Bill Brewer's.

In my discussion of visual anticipations from section 3.2 above, I claimed that anticipations are continuously stirred up in accordance with our movements, which are partly determined by our interests, goals, and visual attention. This feature of my account serves as a remedy for one of Bill Brewer's charges against visual content—let us call it the *selection problem*. The following passage also raises what I will call the *abstraction problem*, which I will discuss below. Brewer asks us to imagine seeing a particular red ball, called "Ball." He then writes:

Ball has colour, shape, size weight, age, cost, and so on. So perception must begin by making a selection amongst all of these. ... Furthermore, and far more importantly for my present purposes, on any given such dimension—colour or shape, say—the specification in experience of a determinate general way that your perception supposedly represents Ball as being requires further crucial abstraction. Supposing that your experience is veridical, it must be determinate to what extent, and in which ways, Ball's actual colour or shape might vary consistent with the truth of the relevant perceptual content. (2011: 79)

AF content is not open to these two worries, the *selection problem* or the *abstraction problem*. First, we need not *select* the properties represented in perception from a long list of possible properties. The content is always determined by our visual anticipations, which are constrained by the movements, context, and interests of the perceiver. The visual anticipations stirred up when we see Ball will always depend on what we are doing when we see it. Are we going to pick it up and throw it to someone? Are we admiring its shiny polish? Is it a piece of clutter interfering with our search for something else? All of these situations would stir up different visual anticipations because they would involve different self-generated movements.

Brewer's second worry in the passage above is what we can call the *abstraction problem*. Visual experience is of particular objects, yet visual content has to do with abstract properties of those objects. My view faces no such problem because I claim that factual properties are always incompletely represented in experience. The abstract property of, say, uniform redness, is a way of intelligibly unifying all of my visual anticipations about the color of the ball. But I never perceive the entire uniform redness of the ball—either as abstract or particular. Instead, my perception of the redness of the ball involves an ongoing process of anticipating changing color experiences. Those anticipations are fulfilled by particular perspectives on the ball. Brewer's charge against orthodox accounts of visual content is that they are committed to the idea that veridical perception "does not consist in the simple presentation to the subject of various constituents of the physical world themselves. Instead, it offers a determinate specification of the general ways such constituents are represented as being in experience" (2011: 80). AF content involves no such generality; instead, it involves anticipations about particular ways objects will appear from particular perspectives. As noted above in my discussion of phenomenalism, those anticipations always depend on the perceiver being intentionally directed toward general factual properties, but I am not thereby committed to the idea that the generality itself shows up in perception. Rather, the generality always goes beyond the content of the actual anticipations. My point here is close to one that has been made by Sean Kelly (2001) in the debate over nonconceptual content. He has pointed out that demonstrative reference to "that color" of an object is actually a kind of abstraction away from the particular color appearance of an object, an appearance that is sensitive to contextual features such as lighting conditions. AF content is meant to capture the particular appearances, including contextual features: I anticipate the way that this particular object will appear to me now if I move in this particular way. Since AF content is not the abstract content of "that color," but rather the content of the way that color will appear to me in a particular context, then Brewer's worry about the generality of content is no threat.

One possible disjunctivist response at this point may be to deny that unfulfilled anticipations are properly regarded as an element of visual experience. The disjunctivist could claim that the visual experience is strictly the single perspectival appearance, minus any anticipations about how appearances would change with movement. This move brings us back to Siegel's thought experiment about the doll from the previous chapter. In making this move, the disjunctivist would deny any kind of perspectival connectedness (PC): the good and the odd experiences of the doll would have no experiential difference. I take it as undeniable that competent perceivers would all report that they expect experiences to change when they move under normal conditions. The point at issue is whether these expectations show up at the level of perceptual experience. Siegel's thought experiment is meant to show that they do, but it remains open to the disjunctivist to deny Siegel's intended results. Denying these results strikes me as wrongheaded. Importantly, AF content is not open to the problems that Travis and others have with propositional content, so there is no real independent motivation for denying perceptual anticipations. Similarly, when disjunctivists talk about the nonperceptual content that is based on perceptual experiences, that content is in the form of a *judgment* that is somehow prompted by our perceptual confrontation with the world. *Perceptual anticipations do the fundamental work of representing that there is a world beyond one's particular perspective, a world about which we can form judgments.* In denying AF content, the disjunctivist would be claiming that we must continuously form the judgment (rather than simply perceive) that there is a visual world beyond our current (and ever-changing) perspective. I do not think it would be unfair to regard such a claim as a gross overintellectualization of visual experience.

Finally, I should make a quick remark about the dialectics here. Readers may notice that my replies to the disjunctivist often rely on the fact that the critical target of disjunctivism is not my own view. Indeed, both I and the disjunctivist reject perceptual propositionalism. What I have tried to show is that one can reject perceptual propositionalism without going so far as the disjunctivist in rejecting perceptual content altogether. AF content, I hope, offers an attractive middle ground between the two extremes currently on offer in the philosophical landscape.

4.4 Four Problems and Three Solutions

So far in this chapter I have presented AF content and some of its rivals. I have made the case that perception has content because anticipations can have degrees of accuracy. Now I am going to show how my claim solves a number of outstanding problems in the philosophy of perception: the speckled hen, skepticism regarding introspection, and phenomenal overflow. In general, these problems are generated due to a simplistic description of perceptual content. With a more precise description, using anticipation and fulfillment, the problems go away. My model does generate what I above referred to as the Davidson/McDowell worry, and I will address this worry as well.

The Speckled Hen—Solved

The puzzle of the speckled hen has been around for decades (Chisholm 1942), and it has seen some attention in more recent literature (Dretske 1993, 2010, Tye 2009, 2010, Nanay 2010). Suppose you take a quick look at a speckled hen in good lighting conditions. The facing side of the hen has 56 speckles. It seems to you as if you see all the speckles on the facing side of the hen, but you do not know precisely how many speckles are on the facing side of the hen. You cannot report that you saw 56 speckles, yet it seems to you as if you have seen all of the speckles. The puzzle is that you saw all 56 speckles, but you cannot report seeing 56 speckles.

Fred Dretske and Michael Tye have taken opposing positions on this puzzle, and Bence Nanay has called something of a stalemate between the two. Both Dretske and Tye agree that we do not see the facing side of the hen as having 56 speckles. But they disagree over whether we see each speckle. Dretske argues that seeing all the speckles amounts to seeing each speckle. Tye maintains that we can see all the speckles without seeing each speckle. Dretske faces the challenge of explaining how we can see each speckle without seeing the facing surface as having 56 speckles. Tye faces the challenge of explaining how we can see all the speckles without seeing each speckle.

Nanay has suggested that there are some cases in which Dretske's analysis fits best, and others in which Tye's analysis is best. He makes his case using Dretske's flock of geese variation on the speckled hen. For instance, when a flock of geese is some distance away, it would be most natural to accept Tye's description: we see all the geese, but not each goose. When the flock is close, and one can take a longer look at it, then it would be natural to accept Dretske's description: we see all the geese, and we see each goose. Nanay concludes that they are both correct and that there is no absolute answer to the puzzle of the speckled hen, or the flock of geese, as it were.

AF content offers a finer level of analysis, a level that focuses on the important issues behind the puzzle: perspective and temporality. One way of paraphrasing Nanay's treatment of the puzzle is to say that perspective and temporality matter for perceptual content. The representational content that one is able to acquire about any visual scene depends crucially on one's perspective and the length of time one is able to explore that scene.

Still, I would like to take Nanay's suggestion even further. According to the model of content that I have been proposing, the puzzle can only be formulated by speaking about perception in an imprecise manner. The source of the imprecision is the fact that objects and visible factual properties can only be *incompletely* perceived; we are constrained to one particular perspective at a time. It is a problematic simplification to claim that one sees all the speckles, or the hen, or the flock of geese. In order to be precise, and avoid problems, one must specify the context, perspective, and duration in which one sees objects or properties. Without such specification, problems such as the speckled hen will arise.

The way in which AF content solves, or simply avoids, the problem of the speckled hen is that it provides an objection to the initial formulation of the problem. We do not simply see all the speckles. Seeing is a process of exploring the facing surface of the hen with a series of saccades, perhaps along with head and bodily movements. The process of exploring the side of the hen involves anticipation and fulfillment. If I saccade into the middle of the patch of 56 speckles, I implicitly anticipate that the indeterminately appearing speckles in my periphery will become clear to me if I saccade toward the edge of the patch of speckles. If I take the time carefully to saccade onto each speckle, then I see each of them. If I quickly scan the side of the hen, then I see all of them, although some of them only peripherally and indeterminately. If I accompany each deliberate saccade with the cognitive activity of counting, then, if things go well, I can report that there are 56 speckles on the facing side of the hen. If one specifies the details of viewing conditions and describes content in terms of anticipation and fulfillment, there is no puzzle.

Skepticism Regarding (a Type of) Introspection—Solved
Eric Schwitzgebel has recently made the case for skepticism about introspection (2008). It may be the case, he suggests, that we are unable to form accurate judgments about the character of our own mental states. Schwitzgebel offers reasons to be skeptical about the character of emotions, dreams, and thought. He also gives reasons to be skeptical about the character of visual perception, which is what I will address here.

There are two lines of reasoning that Schwitzgebel offers in support of skepticism about visual perception. The first line of reasoning is somewhat radical, and will be left aside here. The second line of reasoning, in contrast, can be addressed quite easily using AF content. The first line of reasoning questions the Cartesian tenet that we may be wrong about the objects of perception, but we are not wrong about having the percepts themselves. Schwitzegebel suggests that we entertain the possibility that we are wrong about the nature of our percepts in addition to being wrong about what the percepts represent. We may be dreaming or the subject of a malevolent neuroscientist who causes us to judge that we are having a red sensation, while in fact we are having no such sensation whatsoever. It is not clear to me whether the anticipation-fulfillment model can address such a scenario. On one hand, one might claim that it is impossible to have a fulfillment of a red sensation without actually experiencing a red sensation. The fulfillment just is the having of a red sensation. On the other hand, a skeptic following Schwitzgebel might insist that a fulfillment of a red sensation is not sufficiently unlike a judgment that one is having a red sensation to discharge the skeptical worry. I leave open the skeptical possibility raised by appeals to dreaming or malevolent scientists.

The skepticism raised by Schwitzgebel's second line of reasoning, on the other hand, can be discharged with AF content. The second line of reasoning deals with specific claims about the nature of visual experience; in particular, Schwitzgebel points out anecdotal evidence that many naïve perceivers falsely believe their visual field involves a broad field of clarity. Keeping one's eyes fixed on one spot and shifting one's attention to the visual periphery, as in Dennett's playing card example, can quickly reveal that the area of clarity is quite small. The periphery is indeterminate, but we do not usually notice because we constantly move our eyes. Schwitzgebel elaborates:

> If naïve introspectors are as wrong as many later confess to be about the clarity and stability of visual experience, they're wrong about an *absolutely fundamental and pervasive* aspect of their sensory consciousness. (2008: 256)

His thesis that naïve introspection is unreliable, in the case of vision, looks to have strong support. Schwitzgebel then points out that his argument would work even if naïve introspectors are not wrong after all about the clarity and stability of visual experience; Schwitzgebel entertains the possibility that *he* is "wrong and visual experience is a plenum" (ibid.). This second possibility plays well into a general skeptical framework, but it strikes me as unmotivated.

The skepticism about the initially surprising result is unmotivated, because the initially surprising result finds support in phenomenology as well as empirical science. In chapter 2, I sketched some of this support. Both Husserl and Dennett have noted the phenomenology of peripheral indeterminacy (Husserl 1973b: 340, Rojcewicz trans. 1997: 294, Dennett 1991: 53–54). Besides the phenomenology, there are fairly recent experimental paradigms that provide robust evidence for peripheral indeterminacy (Newton and Eskew 2003, Hansen, Pracejus, and Gegenfurtner 2009, Cohen and Dennett 2011, Freeman and Simoncelli 2011).

Because of the strong phenomenological and empirical evidence, I suspect that Schwitzgebel has given slightly too much to the skeptical view. There are good reasons to think that visual acuity is limited to the small area of the retina corresponding to the fovea, and that we move our eyes in order to compensate for this limitation. One reason why this initially surprising result might be resisted is that it does not fit well with reports of visual content using natural language, reports commonly found in some of the standard literature, as reviewed above.[4] AF content, in contrast, is tailor-made for the frequency of eye movements and the indeterminacy of peripheral vision. Each saccade typically fulfills previous anticipations to various degrees of determinacy. As new areas of the visual field come into view, new anticipations are stirred up.

Before moving on to the next controversy that the anticipation-fulfillment model can settle, I should make a general remark about skepticism regarding the nature of one's own mental states. One point that seems often overlooked is that there is a kind of ignorance of mental states which is not surprising or disturbing at all. I will explain. One way to formulate a kind of skepticism about introspection is to claim that we might not be able to know our own "current conscious experience" (Schwitzgebel 2008: 245, Bayne 2015). There is a sense in which this claim supports a disturbing skepticism, and another sense in which, I think, this claim is plausible and not disturbing at all. On one hand, Schwitzgebel's project seems to be geared toward the first reading, the reading of the claim as a disturbing skepticism. On the other hand, Schwitzgebel does not discuss the way in which the claim can be taken as true. The claim that we might not be able to know our current conscious experience is plausible if "knowledge of current conscious experience" means an exhaustive description of one's conscious experience at a particular instant. The ongoing process of anticipation and fulfillment involves some indeterminacy, which is both spatial and temporal. There is always something indeterminate about experience, something which one cannot quite articulate. Husserl, for one, was clear that particular experiences cannot be fully grasped (Husserl 1976 §44). In sum, although complete ignorance of our current conscious experience may be disturbing, it should not be disturbing, and it is quite plausibly true, that we do not have complete knowledge of our current conscious experience.[5]

Phenomenal Overflow—Solved

George Sperling's well-known experimental paradigm from 1960 has received a flurry of attention recently from philosophers. Most notably, Ned Block has appealed to Sperling's results as a main part of his argument that there is "phenomenal overflow" in conscious experience. That is, Block has argued that there is more to conscious experience than we can access (and report) at any particular time (Block 2007, 2011). First I will describe Sperling's results and then present Block's views along with the views of some of his critics. Then I will show how AF content provides a straightforward way in which to understand Sperling's results, a way which unites various themes in the debate.

T D R

S R N

F Z R

Figure 4.2
An example of one of the stimuli from Sperling (1960).

Sperling (1960) presented his subjects arrays of between 3 and 6 alphanumeric characters in either one, two, or three rows simultaneously, as in figure 4.2. These stimuli were presented for 50 ms. Sperling reports that "When complex stimuli consisting of a number of letters are tachistoscopically presented, observers enigmatically insist that they have seen more than they can remember afterwards, that is, report afterwards" (1960: 1). What adds to the enigma are the experiments in which Sperling introduced a brief tone shortly after the visual stimulus. Subjects were presented with the stimulus for only 50 ms, and then they heard a tone that was either high, medium, or low. The tone served to instruct the subject whether to report the top, middle, or bottom rows. Using the tone to request a partial report resulted in a dramatic increase in accuracy compared to the experiments in which subjects were instructed to report the entire display. In psychology, this work played an important role in theoretical developments regarding iconic memory. In twenty-first-century consciousness studies, it has served as a main source of evidence for Ned Block's theory that phenomenal consciousness "overflows" cognitive access.

Block (2007, 2011) has used Sperling's results, in conjunction with other lines of evidence, to claim that "the machinery of phenomenology is at least somewhat different from the machinery of cognitive accessibility" (2007: 489). The full display, according to Block, appears in phenomenal consciousness, which is why subjects report seeing the entire display. Block goes on to suggest that the working memory system, which enables us to report the elements of the display, does not have the capacity to encode the entire display at once, which is why subjects are unable to report the entire display. The higher capacity of visual phenomenology "overflows" the lower capacity mechanisms of access and reporting. Having made these claims, Block goes on to draw somewhat controversial conclusions about which neural structures constitute visual phenomenology.

As a number of critics have pointed out, Block fails to consider phenomenal indeterminacy in his interpretation of the empirical results (van Gulick 2007, Cohen and Dennett 2011, Stazicker 2011). Block's reply to this charge is to note evidence that the attentional blink is a binary phenomenon, there is no indeterminacy (Block 2007: 533, Sergent and Dehaene 2004). This reply fails for at least two reasons. First, there is no reason to think that a specific, laboratory induced, phenomenon generalizes to all of visual experience. Second, the most robust example of visual indeterminacy can be found in the visual periphery, as discussed above. Block's theory, as far as I know, includes no account of peripheral indeterminacy. I should also mention here that some of the evidence cited by Block has been challenged on methodological grounds (Overgaard et al. 2006).

The objection to Block's claims about "phenomenal overflow" by appealing to indeterminacy seems to me to be exactly right. The reason that naïve subjects claim to see all the characters in the display is that they are, well, naïve. They have not noticed or learned about the ubiquity of visual indeterminacy—few subjects, except some psychologists and some philosophers, tend to know about such things. Still, this objection is lacking. In particular, it generates the question of how phenomenal indeterminacy is compatible with improved performance on the partial report task using the tone as a cue for which row to report. If all the figures in the display are perceived indeterminately, then the partial report should include some indeterminacy as well.

Ian Phillips (2011) deals with the question by pointing out that we can explain the results without drawing conclusions about conscious experience. Perhaps so, but this suggestion does not help with the issue of what we *should* say about the puzzle of conscious perceptual content. James Stazicker comes closer in dealing with the problem by arguing that "the effect of cuing was not to allow access to a subset of the information of which Sperling's subjects were conscious. Rather, the effect was to alter their conscious experience such that some information became more determinate in it" (2011: 169). This response is along the lines of what I want to claim, but I suggest that anticipation and fulfillment offer more details about the process of indeterminate content becoming more determinate.

Sperling's paradigm creates a situation in which the quick flash of the number grid generates anticipations in the subject sufficient for identifying the numbers. Since objects do not typically disappear into thin air, we anticipate that saccading onto a particular number will bring that number into more clarity. Before we can saccade onto each number, the display disappears and our anticipations are disappointed. Crucially, perceptual anticipations are fleeting, and there are always some anticipations that are more determinate than others. The effect of the tone, I suggest, is to stir up the anticipations of the cued row of characters before those anticipations fade away. In this way, those anticipations become more determinate, while the others fade into unreportable indeterminacy. This account of the Sperling results follows Stazicker's account; both of us emphasize perceptual indeterminacy. But my adding the structure of anticipation and fulfillment includes the detail that the indeterminate becomes more determinate as the anticipations that had just been disappointed become stirred up once again with the tone. The subject is able to respond accurately by, in effect, asking herself what she would have anticipated seeing in a particular row.

My reply to the Sperling paradigm also applies to more recent results that have been used by Block (2014) in order to claim that conscious visual perception is "rich" outside of our focal attention. Zohar Bronfman and colleagues (2014) presented subjects with stimuli of 24 letters divided into four rows. Each letter was colored, and the colors in each row were either highly diverse (with hues picked randomly from the color wheel) or less diverse (with a set of hues adjacent to each other on the color wheel). Before each stimulus, a row was cued. After each stimulus, a small box was presented in the area where a letter was located in the cued row. The subject's main task was to name the stimulus letter that was located in the area identified by that box. A second task for the subjects was to report high or low color diversity for either the cued row or the three uncued rows. Subjects performed well on the color-diversity task for both the cued and the uncued rows, with only a small drop in performance for the uncued rows. Also, performance on the color-diversity task did not hinder performance on the primary task of naming the letter, even when the secondary task involved uncued rows. Block draws the following conclusion:

The fact that subjects have almost as much awareness of color diversity in uncued rows as in cued rows suggests awareness of individual colors that are not focally attended above the capacity of visual working memory, supporting the rich view of visual consciousness. (2014: 446)

The view I have been urging is that peripheral visual experience is indeterminate, so I would like to resist Block's interpretation of these results as showing that peripheral vision involves rich conscious perception. The straightforward way to resist this interpretation can be seen if we consider the nature of the task. Note that subjects were not asked to name individual colors in the uncued rows, but only to make the binary judgment whether the rows were diverse with respect to color. Thus, Block's claim that the results suggest "awareness of individual colors" is not warranted. The more precise description of the subjects' ability is to say that they were aware of whether there was diversity of color. This kind of ability is easily accommodated by appeal to indeterminate visual anticipations. In order to complete the second task when it involved uncued rows, subjects need only consider whether visual exploration of those rows would involve anticipations of color diversity or (relative) color uniformity. The content of those anticipations need not be rich, and need not include details about each individual color. Now, Block has framed the debate as an issue of whether or not the information in peripheral vision is conscious (and rich) or unconscious. I recommend a third option: peripheral vision is conscious but indeterminate. The colors in the periphery need not be determinately experienced in order to report accurately whether they are diverse.

The Davidson/McDowell Worry—Unsolved, but not Unsolvable

Although AF content allows one to dismiss a number of problems in the contemporary philosophy of perception, it does raise a well-known epistemological worry: the problem of how perception can justify belief. It is widely accepted that beliefs are a kind of propositional attitude and that we can express beliefs using natural language. If perceptual content has the structure of the ongoing process of anticipation and fulfillment, then it is unclear how it can serve as the justificatory basis for a belief.

Richard Rorty has expressed the worry as follows:

Nothing counts as justification unless by reference to what we already accept, and there is no way to get outside our beliefs and our language so as to find some test other than coherence. (Rorty 1979: 178, cited in Davidson 1983/2001: 141)

Davidson cites this passage approvingly, and is thereby motivated to abandon the attempt to justify beliefs with perception.[6] As is well known, Davidson takes the position that the relationship between perception (or sensation) and belief is causal, rather than justificatory (1983/2001: 143). John McDowell has also taken this theme seriously in his work. Instead of following Davidson's strategy, McDowell has used this worry to motivate his conceptualism about perceptual content (McDowell 1994).[7]

A similar worry surfaces in recent work by Dan Hutto and Erik Myin (2013) under the name of the "intelligible interface" problem. This is the problem of explaining how nonpropositional perceptual content "might intelligibly interface, in virtue of their contents, with other states of mind" (2013: 103). My remarks on ways to handle the Davidson/McDowell worry can also be applied to Hutto and Myin's version of the problem.

Hutto and Myin's main thesis, radical enactive cognition, "denies that basic minds are contentful" (2013: xii). But note that this view also faces the intelligible interface problem. They allow that:

Doubtless some experiences *inspire* conceiving, and even conceiving of things as being a certain way—i.e., of judging them to be thus and so. (2013: 32, emphasis added)

and

Perceptual experiences can *incline* or *prompt* explicitly contentful beliefs and judgments, but they do not, in and of themselves, attribute properties to the world. (2013: 87, emphasis added)

These passages indicate that they are committed to *some* kind of relationship between perceptual experience, on one hand, and beliefs and judgments, on the other. If AF content owes an account of the "interface," then radical enactive cognition owes an account of the way in which perceptual experience can "inspire" or "incline" or "prompt" beliefs and judgments.

I will make three points about the relationship between this epistemological worry and AF content. The first point is that this worry is not an objection to my account. Following Davidson (and perhaps Hutto and Myin), one can simply abandon a justificatory relationship between perception and belief. The second point is that one could take an eliminativist stance with regard to beliefs in order to avoid this worry (Churchland 1981). Of course, this move may lead to more objectionable worries about rationality and language. The final, and perhaps most interesting, point is that one could address this worry by questioning Rorty's initial claim behind it.

Taken as a matter of common sense and everyday discourse, Rorty's suggestion seems obviously true. Beliefs are justified by other beliefs, which can be expressed using natural language. Importantly, though, common sense and everyday discourse need not be the final arbiters of philosophical truth. As it turns out, Husserl devoted much of his philosophical efforts to exploring the relationship between perceptual experience, on one hand, and predicative judgments based on experience, on the other. Some of his earliest writings address the topic (1900/1993 V&VI), and it is the focus of his last major work, *Experience and Judgment*, published posthumously (1948/1973c). Similarly, the question of rational justification was also an important theme in his work (1900/1993 VI, 1976 §24). Although it is not immediately obvious how his work on these topics might feature in a solution to the Davidson/McDowell problem, I would like at least to mention some promising work in this area.

For instance, Michael Barber has shown that some themes central to McDowell's work on perceptual judgments overlap significantly with Husserl's treatment of active and passive synthesis (Barber 2008). It would be off track to enter into the details of Husserl's treatment and of Barber's exegesis, but one quick point is worth mentioning. Barber explains, following Husserl, that indeterminate perceptual content is not conceptualized, but it is *conceptualizable*. That is, further exploration—either motor or attentional—can actively bring the indeterminate into conceptual determinacy. This point is relevant because it raises the possibility of a kind of nonconceptual content that is different from the nonconceptual content that worries McDowell. For McDowell, the concern is that a nonconceptual "bare particular" or "given" is supposed to play a role in rational justification. On Husserl's analysis, as Barber indicates, the nonconceptual is not a bare particular, but, rather, is always conceptualizable (2008: 84; also see Hopp 2010 and Doyon 2011). Bringing Husserl's framework into this recent debate adds an interesting new variation. For Husserl, perception is an ongoing process, with the constant interplay between active predication and passive receptivity. The general structure of this process is one of anticipation and fulfillment. It is not immediately clear that this framework can provide rational justification from a nonpropositional basis, but it does at least show nonconceptual content need not be the bare particular that McDowell finds so epistemologically troublesome.

From a more interdisciplinary perspective, there have been proposals that integrate Husserlian themes with formal models of perception and cognition. Jeffrey Yoshimi (2009) has proposed a "two-systems" approach to Husserl's theory of belief formation. One system is involved in the ongoing synthesis that enables any perceptual experience—similar to the ongoing process of anticipation and fulfillment in the present work—and the second system involves deliberate, active attention toward particular properties of objects. Yoshimi argues that such a two-systems approach fits nicely with dynamical systems models from cognitive science.

In an ambitious development of Husserl's themes from *Experience and Judgment*, Jean Petitot introduces a formal treatment of the way in which perceptual experience can relate to propositional judgments (Petitot 2000). One way of expressing the problem of perceptual states providing rational justification for propositional states, such as beliefs, is as follows. Propositions and rational justification are both properly studied using formal logic, but the domain of logic does not extend to nonpropositional perceptual states. Thus, nonpropositional perceptual states cannot provide rational justification. Petitot's solution to this problem is to use results from category theory, results that unify logic, on one hand, and algebraic geometry and topology, on the other (MacLane & Moerdijk 1992). Petitot treats perceptual states formally by using algebraic geometry and topology, and then he shows how category theory allows the introduction of semantics based on intuitionistic logic. Independent of the success of Yoshimi's or Petitot's particular approaches, these strategies raise an important point. If Rorty's claim is not to be evaluated on the practice of everyday language users, then one challenge to his claim can come from developments at the intersection of formal epistemology and pure mathematics. The epistemological worry about nonpropositional content may not be a worry after all, but rather the beginning of a fruitful research program.

4.5 Summary

The purpose of this chapter was to introduce AF content and make the case that it has advantages over other theories of perceptual content. AF content is motivated by the three general features of visual perception presented in the second chapter. Rival theories of content are not sensitive to these features, and seem to be in tension with them. In the final part of the chapter I have shown how AF content solves three outstanding problems in the philosophy of perception. Departing from perceptual propositionalism does generate the problem of the rational justification of perceptually based beliefs, but I suggest that this problem is not fatal or intractable.

These first several chapters of the book are intended to provide justification for, as well as some details about, the first premise of my Main Argument. The first premise, recall, is this:

(1) *The descriptive premise:* The phenomenology of vision is best described as an ongoing process of anticipation and fulfillment.

In the following three chapters, I make the case for the second premise of the Main Argument. The next chapter covers the empirical psychology of vision, and the chapter after that focuses on the neuroscience of vision.

Part II

5 Some Perceptual Psychology

In this chapter and the following two, I make my case for the second premise of the Main Argument, which, recall, is this:

(2) *The empirical premise*: There are strong empirical reasons to model vision using the general form of anticipation and fulfillment.

In this chapter my goal is to introduce evidence from perceptual psychology in support of (2). In the following two chapters I will add neuroscientific evidence. There is a sense in which this division between psychological and neuroscientific evidence is misleading since a great deal of contemporary research combines the two. Still, the division is helpful as a general organizational principle. The first main point of this chapter is that premise (2) represents a general framework that brings together a number of influential approaches to visual perception. After making this first point, which is mostly historical, I will turn to the main results from perceptual psychology that motivate (2).

These results fall under two basic categories. First, there are the results indicating that vision does not involve a fully detailed experience of the visual environment. Next, there are results indicating that vision is closely connected with action. Together, these two lines of empirical results lead us to premise (2). In the final sections of the chapter, I will address common reasons given for resisting models of visual perception that are similar to (2), and I will explain how my framework offers a synthesis of two existing accounts of visual attention. My goal in this chapter is not to argue that premise (2) displaces particular existing theories in perceptual psychology. Many existing theories complement (2), though perhaps with slight modifications. I will leave the details open. But I should note here that there are some relatively extreme views, discussed below, that I do take to be incompatible with premise (2).

5.1 Various Strands of Support

One of the main claims of this book is that there is a convergence between descriptive and empirical approaches to the general structure of visual perception, as reflected by premises (1) and (2), respectively. There are a number of empirical approaches that can be understood as suggesting something like (2). Of course, there are also kinds of empirical results that have been used as strong resistance against anything like premise (2), and I will turn to those later on in the chapter. There are considerable differences between the various approaches that I will categorize as providing support for (2), but my main goal here is to highlight their similarities. In doing so, I will be casting things at a fairly general level of description. In the following sections of this chapter, more of the details will be filled in.

Perhaps the first empirical scientist to approach perception in a way friendly to premise (2) is Hermann von Helmholtz, who is often considered the father of perceptual psychology. Von Helmholtz famously described perception as "unconscious inference" (1867). Years later Richard Gregory further developed von Helmholtz's line of thought by describing visual perception as a process of "hypothesis testing" (1980). How do these ideas lend support for (2)? The core element in both suggestions is that visual perception is an active process of the mind. One way that I can appropriate the ideas from both Helmholtz and Gregory is to claim that visual perception involves unconscious inferences about how appearances should change as we move. These inferences stir up anticipations, and those anticipations can be understood as hypotheses that are tested through self-generated movement. As those hypotheses are confirmed, as the anticipations are fulfilled, we gain further evidence for our representation of factual properties. Of course, my use of their ideas might depart, in important ways, from what they had intended. But I hope at least to pick up on their common insight that vision is an active process. Besides the general point that vision is a process of hypothesis testing, Gregory also placed great emphasis on the role of background knowledge in visual processing. As Gregory (1997) explains, one powerful source of evidence for such a role can be found in the hollow mask illusion, in which a concave hollow mask appears to be convex from a variety of viewing angles. Since the generation of anticipations should depend partly on the background knowledge of the perceiver—the knowledge that faces are nearly always convex—this kind of evidence supports premise (2).

In addition to Helmholtz and Gregory, I would like to enlist elements of Jerome Bruner's "New Look" psychology in support of (2). In the mid-twentieth century, Bruner and his collaborators ran a series of experiments that revealed fairly strong top-down effects on visual experience, effects that are partly determined by the particular sociocultural background of the perceiver. In perhaps the best-known such experiment, they found that boys from a lower economic background overestimate the size of coins to a greater extent than boys from a privileged background (Bruner and Goodman 1947). In general, Bruner's evidence for top-down effects on visual experience are friendly to premise (2) because visual anticipations are top-down. More specifically, what Bruner's results suggest is that the background and context of the perceiver can play a significant role in the content of visual experience. If the general framework that I am urging is correct, then visual experience should be sensitive to background and context in the way that Bruner's results suggest. In the final chapter of the book, I will revisit this topic as I explore how my Main Argument can be fruitful for work at the intersection of visual perception and social cognition. I should also note here that some of the most influential opposition to (2) is explicitly framed as an antidote to the (over)influence of New Look psychology. I will turn to the opposition below, but now only note that my own intention is to include Bruner's results in a balanced way, without drawing some of the more radical conclusions that have met with strong opposition.

Another important figure in perceptual psychology who can be cited as giving early motivation for (2) is James J. Gibson (1950, 1979). It would be a large undertaking to investigate the details about how Gibson's work might be compatible with my general argument, but I should make a few quick points. There are several main themes in Gibson's work that play a crucial role in the framework I am trying to elucidate. Gibson placed great emphasis on the importance of action for visual perception; he pioneered the idea that visual perception involves repeated sampling of the environment, and he gave us the concept of "affordances." The first two points constitute the main substance of this chapter, and the final point about affordances will reappear in the final chapter.

Gibson is also known for his claim that perception involves directly picking up information from the environment (1979). I will not enter into the details of what this claim might mean, nor will I comment on the influential criticism it has received (Fodor and Pylyshyn 1981). But I will note that his claim about perception being direct may lie behind the contemporary situation in which, put roughly, action-based accounts of vision are in tension with neuroscientific accounts of vision. For example, Alva Noë (2012: 16) has recently expressed what he takes to be a shortcoming of neural-based accounts of perception: "Neurons speak only one language, that of the receptive field. And there is no way to say 'presence in absence' in the receptive field idiom" (2012: 16).[1] One theme of the second part of this book is that there may be no need for such tension. In the following chapter, I will discuss the way in which neural processing can accommodate the worries of sensorimotor theorists, such as understanding extra-classical receptive field properties in terms of visual predictions (see section 6.3). Similarly, as Anil Seth (2014) has argued, predictive-processing approaches in cognitive neuroscience offer a way of understanding the neural implementation of counterfactual sensorimotor contingencies, which can be understood as the counterfactual visual anticipations that I have been describing so far. It may be true that orthodox visual neuroscience fails to address some of the main motivations for the sensorimotor approach, but, as Seth and I argue, there are other theoretical options in neuroscience that fit quite well with sensorimotor themes.[2]

The empirical models that support premise (2) most strongly have been developing with increasing influence in the last couple of decades. A combination of Bayesian inference and predictive processing has emerged as a new and powerful kind of model of visual perception, as well as of other faculties of mind. The idea in its most basic form is that the information processing that underlies visual perception, and other faculties, is both probabilistic and predictive. The brain predicts future sensory input based on prior probabilities. Errors in prediction are minimized either through revisions in the internal model (known as perceptual inference) or through selecting actions that minimize error (known as active inference; Hohwy 2013). In the following chapter, I will present the framework in more detail. I should note that I am only appealing to this framework insofar as it provides support for (2). Many scientists and philosophers are excited that the predictive-processing framework holds the promise of explaining all aspects of intelligent behavior (Clark 2013b, Friston 2013), and here I remain agnostic on this issue.

The agreement between premise (2) and the Bayesian predictive-processing approach can be seen if we understand predictive processing to be a functional-level description of visual anticipation. Along the same lines, as I discuss in the following chapter, we can understand feedback connections in cortex as the neural implementation of both predictive processing and visual anticipation. My intention is that visual anticipation is to be taken as describing the structure of both the personal and subpersonal levels in vision. In other words, I consider empirical models that posit prediction as a main element of visual perception to be direct support for (2), even though those models might not use the term "anticipation."[3]

In addition to predictive processing, these models emphasize Bayesian hierarchical inference for visual perception (and other faculties). On this view, perceptual inference is driven in part by prior probabilities (Clark 2013b). The earlier parts of this book have already given us a straightforward way to understand probabilistic coding in terms of conscious visual experience: visual information in subpersonal probabilistic code is reflected in the *indeterminate* nature of personal-level visual experience. When we are less familiar with a visual environment, we have lower prior probabilities assigned to our best predictions about the way the environment will appear as we explore. Such subpersonal encoding is manifest in experience in the form of indeterminate anticipations, as discussed in previous chapters. Thus, one especially exciting feature of these models is that there is some similarity between the nature of the subpersonal processing, on one hand, and the way in which we consciously see the world, on the other (Madary 2012b, Clark 2013a: 7, Seth 2014).

The main point of this section is to suggest that premise (2) finds support from a number of different major strands in the history of vision science, from Helmholtz to Bayesian predictive processing. Now here is an outline of what I take to be some of the most important evidence for (2) from perceptual psychology.

5.2 Rejecting the Myth of Full Detail

In his book devoted to vision, Zenon Pylyshyn makes the following claim about visual phenomenology:

> As we look around, the phenomenal content of our perception is that of a detailed and relatively stable panorama of objects and shapes laid out in three dimensions. (2003: 4)

I suggest that this claim is false. It expresses a myth about vision to which some philosophers and vision scientists subscribe. We can call it the myth of full detail. In this section, I will give my reasons for rejecting the myth of full detail and I will, along the way, show how the falsity of the myth of full detail is one main motivation for describing vision in terms of anticipation and fulfillment.

The first reason to reject the myth of full detail is one that we have already encountered in the second chapter's discussion of peripheral indeterminacy.[4] We only see visual detail in the center of vision, which makes up only a small fraction of the visual field (Freeman and Simonelli 2011). It is not accurate to claim that we see the details of a panorama of objects laid out before us. Instead, we should say that we have implicit anticipations about how the details of those objects might appear were we to take a better look. Even if we take our time and saccade onto different locations on the surfaces of the objects before us, we still only see the details of the precise locations where we saccade. The rest of the details fade into indeterminacy.

There are other reasons to reject the myth of full detail. Change blindness and inattentional blindness are phenomena that speak strongly against the myth. In the change-blindness paradigm, subjects are presented with a visual image of a natural scene, and then given a quick distractor mask followed by a second image. The second image is identical to the first in all aspects except one, typically major, detail. Then the subject sees another distractor, followed by the first image again, and so on. The experimental paradigm is surprising because most subjects are unable to detect the change between the two images unless allowed to look at the alternating images for quite some time (Rensink, O'Regan, and Clark 1997).

The change-blindness experiments reveal how much of the visual scene is not available in detail for report. We perceive the gist of the visual scene and thereby form implicit anticipations about what kinds of information are available through continuous exploration. When the distractor mask is

Some Perceptual Psychology

presented, anticipations are disappointed. This disappointment, or visual surprise, draws in resources that would ordinarily be available to register disappointed anticipations reflecting the smaller change in the visual scene.[5] One might ask, at this point, why the distractor mask violates anticipations given the fact that, after one cycle, the subject expects the distractor. The straightforward answer is that, as Pylyshyn himself and others have argued forcefully (Fodor 1983, Pylyshyn 1999, 2006), visual content cannot be fully determined top-down by beliefs. One might believe that a visual change is about to occur, but, given the abnormal nature of the change, not visually anticipate it. Visual processing evolved over a great many years in order to help us perceive relevant features of the natural world. It is to be expected that visual stimuli generated using modern technology remain visually surprising despite higher-level expectations.

Fred Dretske (2004) has explored a different way to analyze the change-blindness results. He makes a distinction between seeing objects and seeing facts. To borrow his example, one might see an old friend's cleanly shaven upper lip (an object) without seeing that the old friend is no longer wearing a mustache (a fact). While this distinction might be useful for an analysis of mental content at some level, I would like to resist using it for an analysis of visual content. Rather than trying to explain change blindness by saying that we can have object perception without fact perception, I suggest the simpler explanation that the indeterminacy of visual perception allows for the perception of the gist of complex visual scenes plus select details. (This suggestion receives further support in the following section with my discussion of the task-dependence of saccades.) As explained in the first part of the book, it is important to keep in mind that the perception of factual properties is always incomplete and that each visual fixation is always accompanied by a great deal of peripheral indeterminacy. Dretske's object/fact explanation ignores these two facts in order to treat visual content using natural language, a strategy I rejected in the previous chapter.

Now consider inattentional blindness, which occurs when a stimulus is presented clearly in the subject's field of vision, but the subject fails to be aware of it because she is engaged in another task (Mack and Rock 1998). In perhaps the most well-known demonstration of inattentional blindness, Daniel Simons and Christopher Chabris (1999) instructed subjects to watch a video of a basketball game and to count how many times one of the teams had possession of the ball. During the game, an unexpected event occurs:

either a woman carrying an umbrella or a woman in full gorilla suit walks across the court. Roughly half of their subjects experienced inattentional blindness by failing to notice the unexpected event. These kinds of results speak against the myth of full detail. If the myth of full detail were true, we should expect subjects to see the gorilla, for instance, in detail. Instead, many subjects do not report seeing the gorilla at all.

How can inattentional blindness be explained in terms of anticipation and fulfillment? The task that the subject is performing requires attention, which means, in my terminology, that the task stirs up and increases the determinacy of particular visual anticipations. (I will elaborate on this point below.) With the basketball video, the task requires a fairly precise pattern of saccades in order to track the ball. As we saccade in order to follow the ball, other events in the video will be pushed into our visual periphery. Recall that peripheral anticipation is highly indeterminate due to the nature of peripheral processing (section 2.3). Since peripheral anticipations will be indeterminate, the unexpected event may not violate anticipations to the extent that it will influence the way in which visual anticipations are stirred up; or, in other words, the unexpected event may not capture attention. As a result, subjects may not be aware of the unexpected event. This account of inattentional blindness is compatible with the results that indicate fixation on the unexpected event is neither necessary nor sufficient for consciously perceiving it (Pappas et al. 2005). That is, when the unexpected event is consciously perceived without a direct fixation, peripheral anticipations were disappointed. When the unexpected event is not consciously perceived despite a direct fixation, anticipations within central vision were not disappointed sufficiently to redirect attention or change the subject's representation of the gist of the scene.

In sum, there are at least three reasons to reject the myth of full detail: peripheral indeterminacy, change blindness, and inattentional blindness. These reasons for rejecting of the myth of full detail overlap nicely with my suggestions for solving several outstanding problems in the philosophy of perception above (section 4.4). But if we reject the myth of full detail, then we ought to have an alternative account of how the world seems visually present to us. The best alternative account, I suggest, is premise (2). Vision involves the fulfillment (or disappointment) of previous anticipations along with the generation of new anticipations. This process involves continuous exploration through self-generated movement, which brings

Some Perceptual Psychology   99

us to the second line of evidence for (2), evidence that action and visual perception are tightly related. I review some of this evidence in the following section.

5.3 The Importance of Action

The close connection between vision and action has been explored in both the psychological (Findlay and Gilchrist 2003) and philosophical literature (Hurley 1998, O'Regan and Noë 2001, Noë 2004). In this section I will describe some of the most important evidence that shows a link between vision and action.[6] This evidence includes saccades, selective rearing experiments, and reversing goggles. After covering the basic empirical results, I will discuss how the results support premise (2).

Saccades are ballistic eye movements. I mentioned them in the first chapter, and now I will add some more empirical details discovered using eye-tracking devices.[7] Saccades are extremely fast and frequent; we normally saccade three to four times per second. We can, of course, saccade deliberately, but saccades are not under our conscious direction for the most part. In between saccades, we fixate on locations in the visual scene. Perhaps surprisingly, there is evidence that the eye continues to move during fixations, albeit only slightly. These movements are categorized as either drifts, tremors, or microsaccades. Movements during fixation play an important role that vision scientists are just recently beginning to understand and explore in more detail (Martinez-Conde 2009). If these small movements during fixation are counterbalanced using a device which stabilizes the retinal image, subjects experience fading of the visual scene within seconds (Ditchburn and Ginsborg 1952, Riggs and Ratcliff 1952).

As one might expect, especially when keeping in mind the indeterminacy of peripheral vision (chapter 2), saccades are highly task-dependent. The first, and most well-known, study of the connection between saccade pattern and task was carried out by Alfred Yarbus (1967). In the experiments, Yarbus presented all subjects with the same image, which was a copy of Ilya Rebin's painting, *They Did Not Expect Him*. He then gave the subjects different visual tasks and tracked their eye movements while they performed the tasks. For instance, one task was to estimate the ages of the people in the painting. Another task was to remember the location of the people and objects in the painting (see figure 5.1). The eye tracker showed

1 Free examination.

2 Estimate material circumstances of the family.

3 Give the ages of the people.

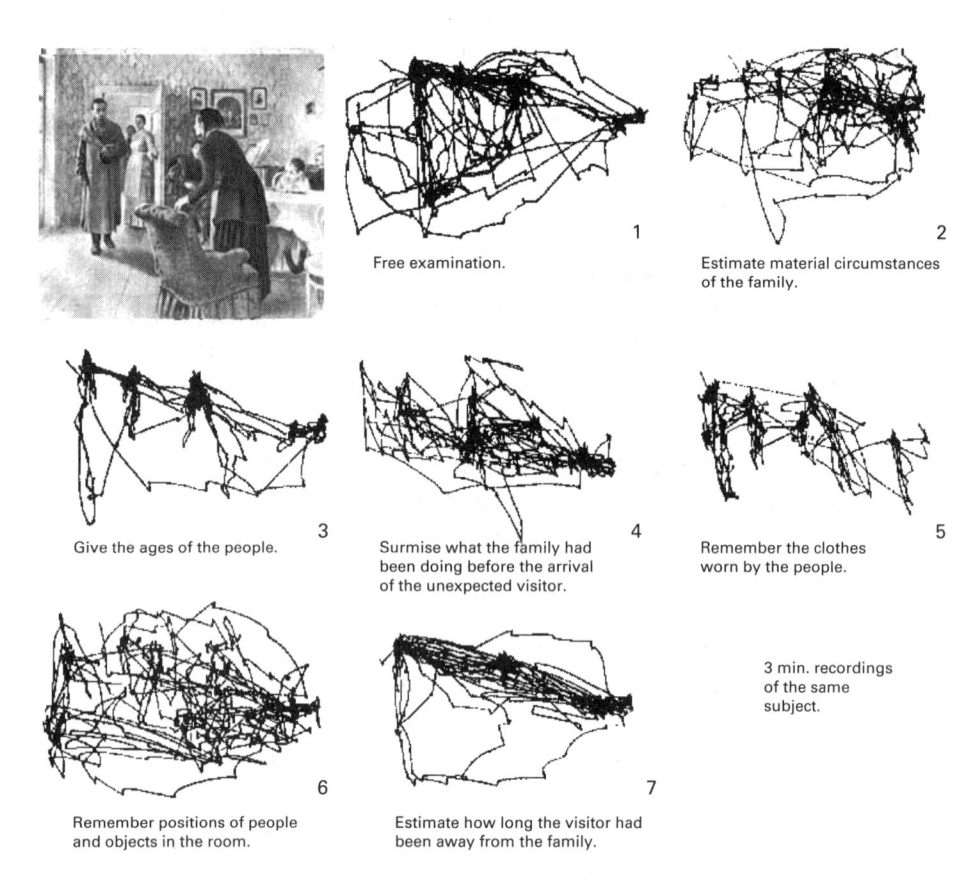

4 Surmise what the family had been doing before the arrival of the unexpected visitor.

5 Remember the clothes worn by the people.

6 Remember positions of people and objects in the room.

7 Estimate how long the visitor had been away from the family.

3 min. recordings of the same subject.

Figure 5.1
Saccade patterns on a stimulus depend on task. From Yarbus (1967).

distinct saccade patterns depending on which task the subject was told to perform.

Advances in technology have allowed for eye-tracking research to move outside of the lab and into more natural environments. Michael Land and colleagues tracked eye movements for subjects who were making tea (1999). They found that nearly every saccade was task-dependent, serving the purposes of locating objects, directing and guiding movements, or checking the state of a task-relevant variable.

The heavy task-dependence of saccade patterns has since been further confirmed by Dana Ballard and his colleagues in both natural and virtual environments (Hayhoe and Ballard 2005). Ballard and colleagues

Some Perceptual Psychology

have made the case that humans use eye movements as a kind of "pointer" for achieving cognitive tasks (Ballard et al. 1997). One motivation for this claim is that such a strategy would be computationally efficient. For instance, instead of programming a hand movement toward one of many possible points in peripersonal space, the nervous system could make use of the simple command to move the hand toward the location where the eyes are focused. Such a strategy would involve a relatively straightforward visuomotor transformation instead of a more computationally demanding strategy.

Here is an example of a study that shows how eye movements can be incorporated into a cognitive task. Ballard and colleagues tracked eye movements for subjects who were asked to arrange a set of colored blocks in order to match a model pattern of blocks. Eye tracking reveals that all subjects looked at the model pattern more than one might expect rather than attempting to store features of the pattern using visual memory (Ballard, Hayhoe, and Pelz 1995). To illustrate, one might expect subjects to complete the task by first looking at the model and remembering the location of a couple of blocks, then looking toward the workspace blocks in order to locate blocks of the remembered colors, and finally looking onto the workspace in order to place the blocks correctly. The eye tracker shows that this strategy is not used. Instead, subjects look back toward the model repeatedly during the task, even while placing the blocks in new positions. The general strategy, it seems, is to store nothing in memory that can be accessed with a quick saccade.

Another relevant line of research involves eye tracking during reading. O'Regan's early work on the topic (1990) serves as one motivation for the development of his sensorimotor approach to perception. Dennett has described his own experience with a program that combines eye tracking with manipulations of text in the visual periphery. The program is able to change the text on the screen without the subject's awareness by predicting the location of each saccade. Dennett illustrates:

What do you see? Just the new word, and with no sense at all of anything having been changed. As you peruse the text on the screen, it seems to you for all the world as stable as if the words were carved in marble, but to another person reading the same text over your shoulder ... the screen is aquiver with changes. ... The effect is overpowering. (1991: 361)

The case of reading provides a nice example of a seamless integration between perception and cognition. And Dennett's example brings out the important fact that visual experience depends very much on the precise location of saccades. We miss out on a lot of detail in the visual world, but we are able to find the details we need due to our ability to anticipate precisely where we need to saccade in order to find what we need.

Moving beyond saccades, another source of evidence for the close connection between action and vision comes from selective rearing experiments. In a famous study published in 1963, Richard Held and Alan Hein paired kittens together using a carousel device. The active kitten was allowed to explore the visual environment while the passive kitten was moved by the carousel as driven by the movements of the active kitten. Thus, the kittens received more or less the same visual stimuli, but the stimuli for the passive kitten did not depend on its own self-generated movements. The active kittens developed normal vision, but the passive kittens showed severe deficits with visual paw placement, the blink response, and the visual cliff test.

A third source of evidence linking action and vision comes from studies on adaptation to goggles that distort vision in some way. For instance, some goggles invert the visual field from left to right or from up to down. The subjects are initially unable to negotiate their environment successfully. After a period of training, adaptation occurs and subjects gain an ability to move around in and manipulate parts of their environment.[8] A variation on the paradigm involves goggles that manipulate color perception. Ivo Kohler (1961) had subjects wear goggles that tint the left side of the visual field blue and the right side of the visual field yellow. When subjects removed the goggles after wearing them continuously for 60 days, they saw the world tinted yellow when they looked to the left, and blue when they looked to the right.[9]

Now consider how these results support premise (2). The basic idea is that we anticipate the sensory consequences of our actions. Let us begin with the frequency and task-dependence of saccades; recall Land et al.'s eye tracking (1999) of subjects preparing a cup of tea. They found that nearly all of the saccades were task-related. A clear way to describe the pattern of saccades is to say that subjects have implicit knowledge about precisely where to look in order to gain information relevant for their goals. If subjects look toward a particular location in order to obtain visual information, then it is most reasonable to say that they anticipate each saccade will reveal information of a particular kind. As explained in the third chapter, this anticipation is ongoing and not a deliberate or discrete act.

The main idea behind (2), that there is evidence for anticipation of the sensory consequences of our actions, fits with selective rearing and distorting goggles in a straightforward manner. The passive kitten was unable to learn how appearances change with self-generated movements during a critical time in development. This inability caused permanent damage. With the distorting goggles, subjects need to relearn how appearances will change as they move. If they are able to practice self-generated movements while wearing the goggles, they eventually adapt by learning the new patterns of appearances. As O'Regan and Noë (2001) would put it, they learn new sensorimotor contingencies.

In this and the previous section of this chapter, I have given some of the empirical results that I take to provide strong support for (2). The first line of results motivates rejecting the myth that we see the visual world in full detail. The evidence suggests that we see detail in central vision along with indeterminacy in the periphery. This evidence supports (2), because the process of anticipation and fulfillment is a process of anticipating the way in which self-generated movement will reveal details of what was previously indeterminate. The second line of evidence shows the close connection between action and visual perception. This second line of evidence supports (2) because the process of anticipation and fulfillment is typically a process that involves self-generated movements of active exploration. My appropriation of this evidence in support of (2) will not meet universal approval. There is a strong tradition in vision science that will resist (2). Now I will face the resistance.

5.4 Facing the Resistance

Perhaps the strongest source of resistance to premise (2) can be found in an approach that treats the science of vision as being properly focused on the purported early vision module, which I will abbreviate as EVM. According to this approach, the science of vision should be concerned with the processing that takes the transduced retinal image as input and then generates a visual representation as output, which may then be sent along to cognitive elements of the mind as input. Proponents of this approach often maintain that such processing is, to use Fodor's term, informationally encapsulated

(1983). That is, information from other faculties of mind, such as cognition, cannot influence the workings of the EVM. Besides Fodor, two of the most important thinkers behind this approach to vision are David Marr (1983) and Zenon Pylyshyn (1999, 2006). If vision is properly understood as the EVM, then much of the evidence cited above can be safely disregarded as nonvisual goings-on. For instance, eye and body movements, as well as top-down influences, can all be quarantined outside of the visual system. If vision does not include these elements, then (2) looks to be highly questionable. After all, EVM involves no anticipation or fulfillment. First I will present the reasons for taking the EVM approach, and then I will explain why these reasons create no difficulty for (2).

But before describing some of the main arguments commonly used in support of the EVM, I should make a quick historical note. One historical motivation for the EVM is to combat what was seen as the overinfluence of Jerome Bruner's "New Look" psychology. As mentioned above, Bruner and his colleagues explored how individual differences might influence visual experience. I have already mentioned his finding that children from underprivileged backgrounds greatly overestimated the size of the coins while children from privileged backgrounds did not. Along the same lines, in another well-known experiment, Bruner and Postman (1949) found that subjects took four times as long to perceive a playing card stimulus that was incongruent with respect to color and suit (red spades, for instance) than a congruent card. These kind of results suggest that there can be strong top-down influences on visual experience, which is to say that the processing of the visual percept is partially determined by information contributed from the perceiver rather than purely by information from the stimulus. Since visual anticipations are top-down, Bruner's results fit nicely with (2).

Now, according to Pylyshyn, Bruner's results and subsequent research along the same lines led, in the 1950s to 1970s, to widespread acceptance of the thesis that perception is continuous with cognition, to what he calls the continuity thesis (2003: chapter 2). Adherents to the continuity thesis would emphasize that a perceiver's beliefs and expectations play an important role in determining perceptual experience. Opponents of the thesis, such as Pylyshyn and Fodor, offer evidence that they take to suggest there is an important element of visual processing that is impenetrable, an element that cannot be influenced by the beliefs and expectations of the perceiver. This element is the EVM.

Some Perceptual Psychology

The purpose of this quick historical detour is to illustrate the way in which two extremes have emerged in this debate over the course of twentieth-century vision science. To put things roughly, one extreme would have it that all visual perception is always already permeated with background cognitive states of the perceiver. The other extreme posits a vision module that can never process information other than what is given from the retinal image. In what follows, I hope to navigate a moderate path between these extremes. On one hand, there is strong evidence that some elements of vision are cognitively impenetrable. On the other hand, the evidence does not require the positing of a strictly feedforward processing module such as the EVM. Now I will turn to the strongest evidence commonly presented for the EVM and show how (2) can accommodate that evidence. If there is no reason to posit the EVM as an extra theoretical entity apart from what I recommend with (2), then, I suggest, let's not.

There are a number of arguments commonly used in support of the idea that vision science should be concerned primarily with the EVM; here I present what I consider to be the four most compelling: perceptual illusions, amodal completion, clinical neuropathology, and a conceptual argument against perceptual feedback.

The most important argument stems from perceptual illusions (Fodor 1983, Pylyshyn 1999). Very many visual illusions cannot be influenced by cognitive states such as beliefs. That is, the visual illusion persists even when subjects have the belief that the percept is illusory. (The Müller-Lyer illusion is the classic example of this phenomenon.) This fact motivates proponents of EVM to treat visual perception as something wholly distinct from cognition.

Another strong motivation for EVM is evidence from what is known as amodal completion. When presented with stimuli that seem to overlap, creating some occlusion, we naturally perceive the occluded surface as continuous: we "complete" the surface. As Gaetano Kanizsa repeatedly pointed out (1969, 1976, 1979, 1985, Kanizsa and Gerbino 1982), our perceptual interpretation of the occluded surfaces does not follow the rules of rational thinking, or, arguably, background knowledge or expectations. Instead, our interpretation of overlapping stimuli seems to follow the internal, impenetrable rules of the EVM. For example, when shown a partially occluded shape flanked by two octagons, it would make sense to see the

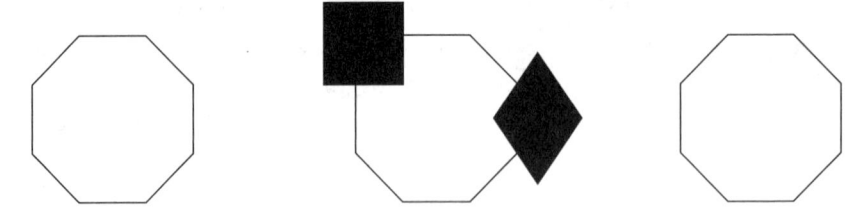

Figure 5.2
Based on Kanizsa's polygons (1985). Instead of perceiving the middle figure as an octagon, we perceive it as an irregular polygon.

partially occluded shape as a third octagon. Rather than as an octagon, we "complete" the shape as an irregular polygon (see figure 5.2).

A third strong point can be made in favor of EVM by appealing to results from clinical neuropathology. Patients with some types of brain damage suffer from a condition known as visual form agnosia (Benson and Greenberg 1969, Farah 1990). Put simply, visual form agnosics have lost the ability visually to recognize objects. This pattern of disability is striking because they have no cognitive deficit, nor do they show abnormalities in basic visual functions such as visual acuity or eye movements. They can visually perceive properties of objects, but they cannot name, match, or copy objects. Pylyshyn has cited visual form agnosia as further evidence in favor of EVM (2006: 71). In patients with visual form agnosia, both the EVM and the stored knowledge about the visual world are intact. It is the ability to integrate the output of the EVM with stored knowledge that seems to be damaged. If stored knowledge and context play a role in low-level sensory processing, then we should not expect this dissociation to be possible. We should expect patients who have trouble with visual recognition always also to show deficits in low-level visual processing. But visual form agnosia offers cases in which the processing of low-level visual features, the processing thought to be carried out by the EVM, appear to be intact while more cognitive, or higher, aspects of visual perception are severely disabled. EVM, it appears, is distinct from cognition.

The final point in favor of EVM is a conceptual one. Fodor has argued that there are good reasons to be skeptical of the role of feedback for perception in general, which would strengthen the case for EVM. His reasoning is as follows:

Some Perceptual Psychology

In short, feedback is effective only to the extent that, *prior* to the analysis of the stimulus, the perceiver knows quite a lot about what the stimulus is going to be like. Whereas, the *point* of perception is, surely, that it lets us find out how the world is even when the world is some way that we *don't* expect it to be. (1983: 67)

Fodor's main point is that feedback will be of limited use for perception, given the fact that one main task of perception is to detect features of the environment even when the perceiver does not expect those features. This point can be raised as a direct challenge to premise (2), and to AF, so it is worth considering carefully.

Now that I have listed what I take to be the four strongest points in favor of EVM, let us consider whether these points threaten (2). The argument for EVM based on perceptual illusions poses no problem, since (2) is not committed to the idea that cognition can determine visual anticipation. As I emphasized in chapter 3, the anticipations that are stirred up in the process of perception are typically not produced deliberately. Still, deliberate, cognitively driven anticipations can occur. For instance, if someone tells you that the hidden side of a teacup has an ugly crack in it, you might deliberately and consciously anticipate seeing a crack as you change perspective. But such anticipations are weak and can be easily canceled out if they are disappointed. That is, if you move and do not see a crack in the cup, you will no longer anticipate seeing a crack in the cup. Likewise, if you watch someone drawing the Müller-Lyer illusion for the first time, and they begin by drawing both horizontal lines without the tails, you might anticipate that you will continue to perceive the lines as being equal in length. When the tails are added, this anticipation is disappointed. Despite the fact that you continue to believe that the lines are equal in length, you cannot force yourself to anticipate perceiving them to be equal in length so long as the tails are attached. To sum up, premise (2) and AF are perfectly compatible with the fact that some belief states do not influence visual perception. These examples might threaten the continuity hypothesis, but (2) does not entail a commitment to the continuity hypothesis.

A similar reply can be used to accommodate the example from Kanizsa. My premise (2) does not imply that the content of visual anticipation should always make sense rationally. Visual anticipations are not stirred up according to the rules of rationality. Another important point to keep in mind with Kanizsa's examples is that they are two-dimensional and highly artificial. In ecologically valid viewing conditions, it is reasonable to think that amodal completion corresponds quite well with the way the world is. Otherwise, we would be continuously surprised as we moved to see previously hidden sides of things. Husserl (1900/1993 VI §10) uses an example of amodal completion from normal viewing conditions in order to illustrate perceptual anticipations. As mentioned in chapter 3, he notes how we implicitly anticipate that the pattern on a rug continues even where the rug is hidden by piece of furniture. We can even modify Husserl's example to address the issue at hand. Imagine you are a guest in someone's living room looking at the pattern of a rug that is partially covered by an armchair. I suggest, following Husserl, that you would implicitly anticipate that the pattern continues under the chair. Now imagine that the host confides in you that he has carefully placed the chair in that position on the rug in order to hide a dark stain. This new piece of information will, I suspect, alter your visual anticipations about how the rug will appear if the chair is removed. If my suspicion is correct, then we have an example of amodal completion being partially determined by nonvisual, or cognitive, information. My example motivates an empirical prediction—a bit of "front-loaded phenomenology," to use Shaun Gallagher's term (2003). One could create, perhaps using virtual reality, an experimental paradigm that builds up visual anticipations in subjects through nonvisual means and then measures reaction times on tasks when those anticipations are fulfilled versus when they are disappointed. I predict longer reaction times and slower task performance when anticipations are disappointed.

The third line of evidence that can be used to support EVM comes from cases of brain damage. Patients with visual form agnosia show a deficit with visual recognition of objects, but show no other cognitive or perceptual deficits. In chapter 7, I will discuss visual form agnosia in more detail when I address the evidence for dual visual systems in humans and other primates. There I will argue that the deficit for visual form agnosics is best described as a deficit in visual processing at a particular spatial scale and temporal frequency. As I explain later on, this description of visual form agnosia accommodates the empirical evidence, and it is much more amenable to (2) than the alternative description in terms of object recognition.

Finally let us consider Fodor's conceptual concern about the usefulness of feedback in perception. His argument has a *prima facie* appeal to it, but there are important considerations that have been omitted. Let us reconstruct it:

Some Perceptual Psychology 109

1. Feedback is only useful when the perceiver already expects what the stimulus will be like.
2. The point of perception is to detect features of the world, even when those features are unexpected.
3. Conclusion: Feedback is of limited usefulness for perception.

The first point to make about the argument is that it is geared toward artificial viewing conditions. In ecologically valid viewing conditions, it is not clear that stimulus detection is the best way to characterize perception. Recall Land and colleague's eye tracking of subjects preparing a cup of tea from earlier in this chapter. The process of visually guided tea preparation is not best described as mere stimulus detection. Instead, it is an active process in which the subject visually accesses information about variables in the environment in a precise and time-sensitive manner. In order to saccade with such precision, subjects need to have some expectations about the visual layout. Thus, in ecologically valid viewing conditions, such as tea preparation, subjects *do* seem to have expectations about what the "stimulus" will be like. If this reasoning is correct, then, according to the first premise, feedback would be helpful in ecologically valid viewing conditions.

Now consider Fodor's second premise. It is true that perception ought to detect features of the environment, especially features that are not expected. How might the visual system achieve this goal? If vision is thought about as the detection of a stimulus on a screen, then modular feedforward processing makes sense. But if we consider facts about the embodiment of the human visual system, such as the fact that there are a great many places one might saccade in the visual environment, and the fact that peripheral visual processing is extremely indeterminate, then a second strategy makes more sense. The second strategy would be to make use of feedback in order to detect anything out of the ordinary in the visual environment. If a feedback prediction, or, to use familiar terms, a visual anticipation,[10] is disappointed, then we can saccade to the part of the environment from which the unexpected visual signal arises and revise our visual representation accordingly. To put this idea perhaps more succinctly, if it is important for vision to detect unexpected features of the visual world, then it makes good sense for the visual system to expect the world to be a particular way; there is no way to detect the unexpected if there are no expectations to begin with. Having reconsidered Fodor's two premises, I urge the conclusion that feedback is of utmost importance for visual perception.

If my suggestions are correct, then Fodor's argument against feedback for perceptual processing can be turned into an argument for feedback, at least in the case of vision. But it is important here to step back and keep in mind the scope of disagreement. All parties can agree that visual perception is not exclusively a matter of feedback. Visual experiences that are created exclusively from internally generated signals are not percepts at all, but are considered to be dreams or hallucinations. The point of contention is whether it makes sense to posit an EVM that is encapsulated from information in the form of feedback, or visual anticipation.

The evidence commonly cited in support of the EVM does not tell against feedback as such. Instead, it shows that there are some kinds of mental states that cannot change the nature of some visual percepts. But this fact is compatible with both feedforward and feedback processing. And this fact is compatible with some kinds of higher-level mental states having an effect on perceptual experience, as is perhaps the case in Bruner's experiments; in some cases of color perception (Hansen et al. 2006, Macpherson 2012); and other cases of purported cognitive penetration of perceptual states (Stokes 2014, Hohwy 2013, chapter 6, Zeimbekis and Raftopoulos, 2015). In summary, there are both empirical reasons (covered in section 5.2) and conceptual reasons (from my appropriation of Fodor's argument above) for thinking that visual perception involves feedback processing. (There is also neurophysiological evidence for massive feedback, which will be presented in the following chapter.) The evidence cited against feedback processing only shows that some kinds of information cannot influence visual processing. It does not pose a problem for all visual feedback. Thus, we can safely maintain (2), that our best empirical models involve anticipatory, or feedback, processing, and add the important qualification that there are constraints on which mental states can have an influence on this processing.

Before moving on to the neuroscientific support for (2), I should address one more issue about the psychological evidence in support of (2). A good bit of recent literature has been devoted to attention, both in psychology and philosophy. In the final section of this chapter, I will explain how (2) fits in relation to some of this work.

Some Perceptual Psychology   111

5.5 Visual Attention

There is a vast body of empirical literature on visual attention, and a rapidly growing body of philosophical literature on the same. It would take us off track to engage with this important topic in full scope and detail, but I should offer some suggestions about how it fits with my Main Argument. In this section of the chapter, I will offer an answer to the following question: If my Main Argument is correct, then how should we understand visual attention? *My answer is that visual attention is, in short, best understood as one way of increasing the determinacy of visual anticipations.* It is important to note that attention is not the *only* way of increasing determinacy. As discussed in chapter 3, increasing familiarity with an object or environment is another way of increasing determinacy. Of course, the process of increasing familiarity would typically involve attention as well.

As I will explain, my account follows two existing accounts of attention in the scientific literature, but recasts them in terms of anticipation and fulfillment. First I will explain how my answer captures our pre-theoretical notion of visual attention. Then I will introduce the evidence for the two existing theories in the scientific literature: the pre-motor theory of visual attention and the understanding of attention as precision optimization. I explain how these two theories of attention converge in my description of visual attention as an increase in the determinacy of anticipations. Then I explain how this hybrid account of visual attention can accommodate some of the most well-known empirical results in this area. I finish this section by addressing possible objections.

As the reader may recall from the third chapter, I suggest that visual anticipations are continuously stirred up in the process of perception. What we anticipate is partly determined by our context and self-generated movements. This process of stirring up visual anticipations goes far in accommodating pre-theoretic intuitions about the nature of attention. One general characterization of attention is the way in which the mind focuses on some mental content instead of other possible mental content. If we apply this characterization to visual attention, then we can describe visual attention as the way in which we focus on, or attend to, some aspect of the visual world. I suggest that attending to some aspect of the visual world just is the same thing as anticipating a more determinate content. For example, if you choose to attend to the pattern on a teacup, your doing so will stir up increasingly determinate anticipations about how the pattern should appear as you access it through saccades and other self-generated movements. This example would be an instance of endogenous attention. For an example of exogenous attention, imagine you are out for an evening stroll and the blinking light of a descending airplane catches your eye. In this case, the blinking light in your peripheral vision violates your implicit anticipation of a still evening sky. Immediately, anticipations are stirred up that saccading toward the location of the blinking light will allow you to see something determinate; in particular, you will see the familiar sight of a descending plane. This account also works for sudden flashes that catch our visual attention and then disappear. A sudden bright flash in the periphery would typically violate general visual anticipations, unless, for instance, one is in a war zone or watching a fireworks display. This violation immediately stirs up anticipations about what might be located in the region of the flash. If we saccade to the location of the flash, we visually anticipate seeing something determinate, some remnant of the cause of the unusual occurrence. Even when no such remnant is visible, the anticipations have been stirred up, and the determinacy increased, nonetheless.

The general proposal is that visual attention is an increase in the determinacy of stirred-up visual anticipations. I have tried to motivate this claim with some quick phenomenological descriptions of visual attention. But the point of this chapter is to show how empirical results support premise (2). I hope to show this by introducing two existing accounts of attention and explaining how they combine to support (2). The first one is known as the "pre-motor," or sometimes simply "motor," theory of visual attention. The second account comes from Bayesian predictive processing and treats visual attention as precision optimization in hierarchical perceptual inference. I will begin with the pre-motor theory.

The pre-motor theory of attention has roots in Alexander Bain's work in the late nineteenth century (1888, Mole 2011). Bain proposed that the mechanisms driving attention are the same mechanisms that drive motor movements. Purely attentional acts occur when the actual motor movement is inhibited. If we consider the fact that *visual* attention typically corresponds with actual motor movements—with saccades—then we see that the pre-motor theory is something of a natural fit with visual attention. That is, the same neural areas that program saccades could determine visual attention at the same time. When visual attention correlates with saccades, when we look where we attend, when attention is *overt*, then the pre-motor theory looks to fit. Of course, *covert* visual attention is also possible; that is, we are capable of attending to a visual feature that is in the periphery without saccading toward it. When visual attention is covert, the pre-motor theory holds that the same mechanism drives both attention and a saccade toward the attended location, but that the saccade is inhibited at a later stage. In addition to psychophysical evidence in support of the pre-motor theory based on saccade trajectories (Rizzolatti et al. 1987), there is also growing neuroscientific evidence in the past decade. Primate saccades are thought to be driven mostly by an area of prefrontal cortex known as the frontal eye fields. Experiments show that it is possible to microstimulate this area in a way that is not strong enough to cause a saccade, but that seems to enhance neural processing corresponding to the area toward which the saccade would have occurred (Moore and Armstrong 2003, Moore, Armstrong, and Fallah 2003, Armstrong and Moore 2007). These results suggest that both saccades and visual attention are driven by the same neural substrate, by frontal eye fields.

The second understanding of visual attention that can be adopted for my framework comes from Bayesian predictive processing, mentioned above. The predictive-processing framework suggests that sensory input is continuously predicted by an internal generative model. If the prediction is not correct, an error signal propagates upward through levels of the hierarchy until the internal model is revised in a way that accommodates the error signal.

Here is how attention enters into this theory. Since perception occurs in different environments, and these different environments introduce various levels of noise in the perceptual input, the system should have a way to discriminate which error signals are more precise than others. Not all error signals are of equal importance to the system. Error signals generated in, for instance, an extremely noisy environment should not be as influential, not carry the same weight, for revising the internal model as error signals in a more perceptually optimal environment. Thus, the system should have the ability to discriminate more from less precise error signals. The suggestion, then, is that attention should be understood as precision optimization (Hohwy 2012, 2013). Put roughly, when we attend to a stimulus, the weight given to any prediction error should be increased.

Both the pre-motor theory of visual attention and the predictive-processing account of attention as precision optimization are able to accommodate a range of empirical evidence on attention. My own description of visual attention as a way of increasing the determinacy of anticipations is intended as a way of incorporating both of these theories of attention into one framework. In vision, the best way typically to increase the determinacy of anticipation is to saccade toward the area of interest. If the pre-motor theory is correct, and visual attention is driven by saccade generation, and the generation of a saccade stirs up increasingly determinate anticipations, then the pre-motor theory falls in nicely with my description of visual attention. The case is similar with precision optimization. If we understand predictive signals from the generative model as visual anticipations, and we understand an increase in determinacy as an increase in precision, then the predictive-processing account of attention shares a strong similarity with my account of attention. Of course, determinacy is not the same as precision. Determinacy, as I have used it, is a way of describing a phenomenological feature of visual experience. Precision, as predictive-processing theorists have used it, is a second-order statistic in the Bayesian model. My suggestion, then, is that precision optimization, as a functional-level description of neural processing, is experienced as changes in determinacy of anticipation by the subject. This discussion brings us close to difficult issues surrounding the relationship between subpersonal description and personal-level experience. I will discuss these issues in more detail below in chapter 8. For the rest of this section, I will describe how my account of attention as an increase in determinacy of visual anticipation—an account that is meant to include both the pre-motor and the predictive-processing accounts of attention—can explain some of the best-known empirical results on visual attention.

So far, and following the pre-motor theory, my emphasis has been on overt attention. Although overt attention is the norm, many of the psychological studies have been concerned with covert attention. Perhaps the most significant empirical studies on covert visual attention follow a paradigm developed by Michael Posner and his colleagues. Subjects are instructed to focus on a fixation point, and then a cue is presented that is supposed to indicate to the subject where a subsequent stimulus will appear (see Posner, Snyder, and Davidson 1980, for example). The attention is covert because subjects visually attend to a location other than the fixation point. On most of the trials, the cue is accurate. But on some trials, the cue indicates the wrong position for the subsequent stimulus. Posner also ran trials in which there was no cue, but just the stimulus. They measured the subject's reaction times to the stimulus under these three different conditions. As one might expect, subjects were fastest when the cue was accurate and slowest when the cue was not accurate.

Posner's results can be described using anticipation and fulfillment in a straightforward manner. The cue stirs up increasingly determinate anticipations about where the stimulus will appear. The act of attending to the location indicated by the cue can be understood in terms of determinate anticipations. The fulfillment of those anticipations facilitates a faster reaction time. Disappointment of the anticipations hinders reaction time because more determinate anticipations about the cued location are unfulfilled (when no stimulus appear there) and less determinate anticipations about the uncued location are also unfulfilled (when the stimulus appears there unexpectedly).

In addition to Posner's paradigm, the understanding of attention as increasing the determinacy of anticipations can account for some of the most important psychophysical results on visual attention. These include visual pop-out, the attentional blink, and inattentional blindness.

The phenomenon of visual pop-out is one of the main motivations behind the feature-integration theory of attention most famously developed by Anne Triesman (Triesman and Gelade 1980, Triesman 1986). Visual pop-out occurs when subjects exhibit a fast reaction time on visual searches for a stimulus amid distractors. The stimulus pops out because it is salient relative to the distractors, as is the F in figure 5.3. According to the feature-integration theory, some visual features of the stimulus—such as color and orientation—are processed automatically and in parallel, while the perception of objects, which involves the integration of these various features, requires focal attention. It is attention that glues the various features together into individual objects. Although I will not engage with the details of feature-integration theory,[11] I do suggest that visual pop-out can be accommodated easily by appeal to anticipation and fulfillment. Elements pop out of a visual scene because they violate the anticipation that patterns and surfaces are visually uniform. Pop-out elements violate the pattern; they generate an error signal, to use the terminology of predictive processing (Spratling 2012). When visual anticipations are disappointed,

```
T T T T T T T T
T T T T T T F T T
T T T T T T T T
T T T T T T T T
```

Figure 5.3
A visual pop-out stimulus. The F is salient against the distractors.

new anticipations with increasing determinacy are immediately stirred up as we revise our visual interpretation of the world. Again in this case, attention is the stirring up of increasingly determinate visual anticipations. If this account of the pop-out effect is correct, then one might expect adaptation (temporal context) effects on visual salience. That is, if the perception of salient stimuli involves visual anticipation, then visual adaptation may give time for visual anticipations to become more accurate, which should thereby make a difference for subjects' performance in visual detection tasks. In support of this expectation, there is evidence that exposure to particular kinds of pre-stimuli, such as colors and gratings, brings about visual adaptation that decreases search time and changes saccade patterns for search tasks requiring discrimination of color (McDermott et al. 2010) and orientation (Wissig and Kohn 2012), respectively.

As a second example of anticipation and fulfillment accommodating important psychophysical work on attention, consider the attentional blink (Raymond, Shapiro, and Arnell 1992). Subjects are presented with a rapid series of stimuli and instructed to indicate when they have detected, typically, two particular stimuli. When the second target stimulus is presented shortly after the first, within about 300 ms, subjects are poor at detecting the second stimulus. This paradigm fits with the framework of anticipation and fulfillment by revealing that there might be temporal constraints on the stirring up of determinate anticipations. The detection of the first stimulus fulfills anticipations. The fulfillment is then followed by the stirring up of anticipations for the second stimulus. Within the 300 ms window, there is not enough time for sufficiently determinate anticipations of the second stimulus to be stirred up following the fulfillment of the first stimulus. Without sufficiently determinate anticipations stirred up, the second stimulus goes undetected. Research also shows how other contextual effects, such as nonvisual tasks, can enhance performance on the visual attentional blink (Olivers and Nieuwenhuis 2005). This change in performance could be described as occurring due to a contextual modulation of the process of anticipation and fulfillment in the visual modality.

The final example of important psychophysical results on visual attention is inattentional blindness. The way in which inattentional blindness can be explained with the process of anticipation and fulfillment has been covered above in section 5.2.

5.6 Objections and Replies

In the remaining part of this section, I would like to address two possible objections about what I have claimed with regard to attention and premise (2). These objections have to do with subliminal perception and the relevance of my Main Argument for empirical psychology.

The first objection is that I have made no room for well-known behavioral effects of stimuli that have behavioral effects but are not consciously perceived. It may not be immediately clear how unconscious priming (Mattler 2003), for example, might fit with the framework of anticipation and fulfillment. My reply to this objection is that the ongoing process of anticipation and fulfillment characterizes both conscious and unconscious perceptual content. Some anticipations can be fulfilled without conscious awareness. In general, the less determinate anticipations are the least conscious anticipations. Since indeterminacy comes in degrees, it may be most helpful to understand visual awareness also as admitting of degrees. This line of thinking brings up an important, but often overlooked, conceptual point about unconscious visual processing: when a stimulus has an influence on behavior but is not consciously perceived, the mental processing of that stimulus may be highly indeterminate.[12] In other words, the content of unconscious percepts might differ from the content of conscious percepts of the exact same stimulus because there is a difference in the level of determinacy in the visual anticipations. Although I will not pursue this idea further here, I would like to note that this way of thinking about unconscious perceptual processing marks a departure from standard approaches in the consciousness literature, approaches that assume that the difference between conscious and unconscious representations is not a difference in content. This assumption can be found, for instance, in higher-order thought theories (Gennaro 2004) as well as Prinz's intermediate-level theory (2012).

The second objection is that my description of this evidence in terms of anticipation and fulfillment fails to explain *how* we perceive the world. That is, I have said little about the details of the mechanisms that enable visual perception. Although some implementational details will be covered in the next chapter, this objection is fair, since I do leave out details. In exchange for details, I have offered a thesis about the general structure of the processing that enables visual perception, and I have explained how this general structure can accommodate some of the most important empirical results from human visual psychophysics. If what I am suggesting is along the right lines, then there are particular questions about the details that ought to be empirically tractable. For example, it seems clear that visual anticipation is not entirely penetrable by cognitive states. Nonetheless, visual anticipation must be driven by some kind of stored background knowledge. One interesting question, then, is what kinds of information can be put to use in the stirring up of visual anticipations. Are there contexts in which some information can drive visual anticipation and others in which it cannot? Relatedly, are there hysteresis effects in the generation of visual anticipations? These and similar questions follow from the central claims of this book, and they can be investigated using familiar methods from perceptual psychology.

5.7 Summary

The main goal of this chapter is to make the case for premise (2) based on perceptual psychology. I have covered some of the historical roots of (2), some of the main lines of psychological evidence, responded to competing ways of thinking about vision, and addressed the question of visual attention.

6 The Active Brain

Now, I turn to the neuroscientific evidence for premise (2).[1] In this chapter, my goal is to defend the claim that there are strong neuroscientific reasons to model vision using the general form of anticipation and fulfillment. I will defend this claim in several stages. First, I will make some remarks about the ongoing endogenous dynamics of the brain. These general remarks serve as a primary motivation for my claim that the brain is essentially an active, rather than a reactive, system. The active nature of the brain is important because the ongoing generation of visual anticipations is a kind of self-driven activity. The second stage of support for the neuroscientific version of (2) will focus on results having to do with the anatomy and neurophysiology of the brain; it will focus on the massive feedback connectivity. In the third part of the chapter, I will explore some of the theoretical options that accommodate the facts emphasized in the first two sections. Although some of the details differ in important ways, all of these options suggest the general structure of anticipation and fulfillment. I close the chapter with a summary and some general remarks.

6.1 Ongoing Cortical Dynamics

Let us begin with some features of all living organisms. As Aristotle first emphasized in *On the Soul*, all living creatures have what he called the vegetative soul (see Book II, part 1, for example). Aristotle's claim was biologically motivated—it does not refer to what many people today think of as a soul in a religious context. In today's terminology, we can paraphrase Aristotle to say that all living creatures have a metabolism: We continuously exchange matter and energy with our environment. This continuous exchange of matter and energy enables all living things to maintain their status as being far from thermodynamic equilibrium with the surrounding environment (Haynie 2008). Maintaining this status entails creating and continuously reproducing a boundary between the organism and the rest of the world. Mark Bickhard (2002) has used the example of a candle flame in order to illustrate this property of living things. The flame's existence depends on it being far from thermodynamic equilibrium, and it maintains this status through a continuous exchange of matter and energy. All living things are a bit like that candle flame. Similarly, the enactivism of Francisco Varela and his collaborators places great emphasis on this process of self-maintenance, referring to it as "autopoiesis" (Maturana and Varela 1980, Varela, Thompson, and Rosch 1991, Thompson 2007). This feature of living organisms plays a central role in Bickhard's work as well as in enactivism, and it has recently been appropriated by Karl Friston into his theoretical framework (2013, Clark 2015a). But my mention of this property of living organisms is for a less ambitious purpose. I merely wish to make a point about the brain.

In keeping far from thermodynamic equilibrium, our bodies burn up a lot of metabolic energy. Importantly, a disproportionate amount of this energy is used by the brain (Raichle 2010). This energy is used, in part, to fuel the brain's own endogenous ongoing dynamics; changes in task usually change metabolic cortical activity by only about 5% (ibid.). The brain is intrinsically active, and it maintains its activity by consuming more than its share of metabolic energy. There are a number of lines of empirical evidence that reveal various aspects of ongoing cortical dynamics, evidence from individual neurons, local field potential, and brain imaging. After providing some of this evidence, I will return to a discussion of how the evidence can be brought to support premise (2).

But before entering into the evidence, I should note that the main point of this section of the book has been made with great historical detail in a recent article by William Bechtel.[2] The thesis of his article is as follows:

The evidence for the endogenously active perspective on the mind-brain is extremely compelling and ... in light of it cognitive researchers should fundamentally reconceive their conceptions of cognition and cognitive architectures to incorporate and recognize the significance of endogenous activity. (2013)

This second section of the book takes up Bechtel's suggestion for the case of human visual perception.

We can begin to consider the ongoing endogenous activity of the brain at the level of the individual neuron. Most typical descriptions of cortical neurons present the neuron's function to be one of transmitting a signal.[3] On this standard description, signals are received as input to the dendrites and then sent as outputs from the axons. The neuron "fires," sending an all-or-none signal from the axon, when the difference in electrical potential inside and outside the cell membrane reaches a particular threshold. This standard description gives one the impression that neurons are reactive, that they are passive transmitters of signals from one place to the next.

This standard account of neuronal firing is helpful as a first, basic description of an extremely complex phenomenon. But the standard account is also incomplete in important ways. A number of different empirical findings throughout the twentieth century are not compatible with the standard conception of the neuron as a mere transmitter. In the early twentieth century, isolated crustacean neurons were found to exhibit variable responses to the same input stimulus (Fessard 1936, Arvanitaki 1939, Hodgkin 1948). Subsequent research discovered neurons with intrinsic oscillatory activity, so-called *pacemaker* neurons (Alving 1968), as well as *resonator* neurons, so called by Rodolfo Llinás because they "respond preferentially to input at specific frequencies" (1988: n17). In order further to investigate these results, contemporary work in computational neuroscience involves mathematical models that incorporate the ongoing internal dynamics of different kinds of neurons.[4] One fascinating question, of course, is what this property of neurons might mean for cognitive neuroscience. Llinás himself has proposed an initial way of interpreting what these findings might mean for mental processing:

In principle, one may propose that intrinsic electroresponsiveness generates internal computational states that serve as the reference frame, or context, for incoming information. (1988: 1661)

Subsequent research has attempted to explore the role of ongoing dynamics for the neural processing of perceptual stimuli.

Given the intrinsic dynamics of neural activity, it is natural to expect variation in the way in which sensory neurons respond to the same stimulus. Such variation has long been observed using *in vivo* single-cell recording (Schiller, Finlay, and Volman 1976, Vogels, Spileers, and Orban 1989, Snowden, Treue, and Andersen 1992, Softky and Koch 1993). The standard way to deal with this phenomenon is to regard the variation as mere noise in the system, and to eliminate the noise using statistical techniques (Tolhurst, Movshon, and Dean 1983, for example). This strategy follows the Nobel Prize–winning results of David Hubel and Torstel Wiesel (1959), in which they discovered receptive field properties of cortical neurons in the cat. If proper neuronal function is strictly to respond preferentially to a stimulus in their receptive field, then any deviation from that response can be regarded as a result of noise in the system. An alternative approach would be to explore how ongoing dynamics might systematically influence variation in neural response. Taking up this alternative strategy, Amos Arieli and colleagues (1996) recorded cortical activity in anesthetized cats using optical imaging (voltage-sensitive dye), local field potential,[5] and single-cell recordings. By recording the ongoing dynamics in cortex prestimulus, they were able to predict the way in which the individual cells would respond to the stimulus when presented. This result, they suggest, challenges the received view that variation in neural response should be attributed to noise. Instead, and following Llinás's suggestion above, they claim that ongoing dynamics might underlie contextual influence on perceptual processing, as well as influence behavioral response and conscious experience (1996: 1870).[6]

Besides neural and extraneural recordings, there is also evidence for large-scale ongoing cortical dynamics from brain-imaging studies. Using magnetic resonance imaging and positron emission tomography, scientists are able to detect activity in various regions of the brain, and a great deal of cognitive neuroscience involves using these techniques in conjunction with particular tasks. One surprising find within this field of research is that there seem to be areas of the brain that become more active during non-task activity. In a widely cited article, Marcus Raichle and colleagues (2001) referred to these brain regions as the "default mode network" (for a more recent overview of the literature, see Raichle and Snyder 2007). The nature and function of such a network is currently a central matter of debate in cognitive neuroscience.[7] Without entering into this debate, I only wish to make the point that the evidence cited in support of the default mode is more evidence for the general claim that there is intrinsic ongoing activity in cortex, and it is plausible that this activity plays a meaningful role in our mental lives.

So far in this section I have reviewed some of the evidence showing that the brain is intrinsically active, rather than reactive. If the brain is intrinsically active, then it is reasonable to maintain, following Bechtel and others, that this property is reflected in cognitive functioning. The model of anticipation and fulfillment for visual perception is a model that is partly motivated by the fact that the brain shows ongoing intrinsic activity. I have suggested that visual anticipations are ongoing, have various degrees of determinacy, and can be context-sensitive (reflecting, for example, familiarity with one's environment). These features might be expected from the kind of neural dynamics described above. In addition, competing models of vision, models that are feedforward and context-insensitive, suffer from an uneasy tension with the ongoing intrinsic dynamics of the brain.

6.2 Neural Feedback

In the previous section I made the case, based on three kinds of empirical evidence, that brain activity involves ongoing intrinsic dynamics, and that these dynamics might be relevant for understanding perceptual processing. In this section I would like to discuss neural feedback connections in the cortex and begin to make the case that these connections implement anticipatory processing. In the following section, I will further develop this case by surveying some of the existing theoretical options that fit with the claim that feedback connections implement anticipatory (or predictive) processing.

It is well known that there is widespread interconnectivity between regions of cortex. In a landmark study of this topic, Daniel Felleman and David van Essen (1991) created a neuroanatomical map of the visual areas in macaque cortex. They described the results as revealing a hierarchy of processing, in which different regions seem to be dedicated to increasingly complex properties of the visual stimulus. Importantly, though, they found that the connectivity does not suggest a strictly serial or feedforward sort of processing of information that one might expect based on standard theoretical interpretations of Hubel and Wiesel's results. That is, they found strong *reciprocal* connectivity—both feedforward and feedback connections—between regions. Subsequent work on this topic continues to explore the details of the massive feedback connectivity in the visual brain (Peters, Bayne and Budd 1994, van Essen 2004), even indicating feedback connections where there are no feedforward connections between cortical regions (Rockland and van Hoesen 1994).

All parties seem to agree that there are more feedback connections than feedforward connections in the visual areas of primate cortex. This anatomical fact alone could be taken as a first indication that feedback plays an important functional role (Bar 2007: 283). But there is still uncertainty about the functional role of all of these feedback connections. In an influential article, Francis Crick and Cristoph Koch (1998: 246) made a distinction between driving and modulating inputs to neurons. Driving inputs are alone capable of making the neuron that is receiving the input fire more strongly. Modulating inputs are not capable of doing so, and can only modulate the firing of the neuron. This distinction allows Crick and Koch to formulate the "no-strong-loops hypothesis," which is the claim that reciprocal connections cannot be both of a driving nature. One motivation for this hypothesis is that strong loops, reciprocal driving connections, would lead to uncontrolled oscillations, as in epileptic seizures.

The distinction between driving and modulating neural inputs enables the preservation of the (once) orthodox feedforward understanding of visual cortical processing, which follows Hubel and Wiesel's framework, as well as the framework of Marr, Pylyshyn, Fodor, J. Prinz, and others. The classical receptive field properties discovered by Hubel and Wiesel can be said to be caused by driving feedforward inputs. The massive amount of feedback connections can be described as merely modulating.[8] On this view, the feedforward connections are easily the most important, despite the fact that they are far outnumbered by the feedback connections (Sherman and Guillory 2002).

An important issue, then, is whether the evidence supports the hypothesis that feedback connections are merely modulating. While many earlier studies have claimed to find evidence that feedback in mammalian visual areas is merely modulating (Martinez-Conde et al. 1999), recent work has found evidence of *both* driving and modulating feedback connections in rat auditory (Covic and Sherman 2011) and visual areas (De Pasquale and Sherman 2011). The terrain is complex because studies on this topic have used different techniques on different species of mammal. There are a great deal of open questions in this area, including the question of whether the distinction itself between modulating and driving inputs is too simple (Crick and Koch 1999: 249). It is also worth adding here that a "merely" modulating role for feedback neural connections may not be incompatible with the suggestion that these connections involve some kind of anticipatory or predictive feedback. For instance, in the previous chapter I explored a possible connection between visual anticipation and visual attention. Even if the feedback is not driving, it plausibly plays a major role in attention (Macknik and Martinez-Conde 2009).

My own sympathies, of course, lie with an understanding of feedback connections as playing a major functional role, enabling anticipatory or predictive processing.[9] Such a view is a straightforward way to accommodate the massive feedback connectivity in cortex, with the important caveat that more connections do not necessarily entail a greater functional role. This view also provides an implementation-level description of the kind of information-processing architecture proposed in the previous chapter. Finally, although there is uncertainty about the role of neural feedback from detailed neurophysiological studies, there is some motivation for treating neural feedback as anticipatory from a theoretical perspective. Let us now consider such a perspective.

6.3 Theoretical Options

So far, I have presented evidence for two basic features of the brain: ongoing intrinsic dynamics and massive feedback connections. Now I turn to theoretical approaches to brain functioning that incorporate these features. Pioneering work by Llinás (mentioned above, 2001) and Walter Freeman (1999) represents some of the earliest attempts to account for human perception, action, and cognition in a way that is sensitive to these features of the brain. In the last few years, these properties of the brain are being increasingly incorporated into mainstream cognitive neuroscience. In particular, the predictive-processing approach to cognition, mentioned in the previous chapter, offers a framework that brings together many different lines of evidence in a powerful and elegant manner. First I will introduce the general principles at work in the predictive-processing framework. Then I will turn to some of the most promising ways in which this framework has been applied to visual neuroscience.

One of the core features of the predictive-processing framework is the generative model. Contemporary generative models have been developed from research at the intersection of computational neuroscience and machine learning; they were developed with the goal of improving upon some shortcomings with standard neural networks (Hinton et al. 1995).

Prior to the development of the generative model, the primary way to train (or teach) neural networks was through a process known as backpropagation, which requires adjusting the weights within the network during learning in order to bring about the desired output. But there are two shortcomings with backpropagation. First, it requires a teacher to specify the correct or desired output. Second, it requires a way of transmitting the appropriate weight change to each connection in the network, which may be neurophysiologically implausible (Hinton 2006).

In developing generative models, Geoffrey Hinton and colleagues introduced an alternative to backpropagation learning.[10] The key insight was to design the model such that it is capable of generating its own data. Very basically, the main idea is that the top-down connections generate predictions about the incoming bottom-up sensory information. The only relevant information entering and propagating through the system, then, is the information that deviates from what is predicted, or the error signal. By treating only the error signal as the incoming sensory information, the system saves enormously on computational burden (Clark 2013b: 2–3). The basic operation of the brain, on this view, can be summed up as the process of minimizing prediction error. The system attempts to "explain away" prediction error through hierarchical Bayesian inference. When the system engages in what I described as perceptual inference in the previous chapter, it is altering the generative model. The general principle driving both perceptual inference and active inference is the effort to minimize free energy in the brain (Friston, Daunizeau, and Kiebel 2009, Hohwy 2013).

Note that, until now, I have been describing neural connections in terms of feedforward and feedback. The predictive-processing approach to neural processing effectively flips that description around (Logothetis 2008: 872). In other words, the information that is transduced at the sensory periphery and processed upward through the hierarchy—what would typically be described as feedforward information—becomes a kind of feedback for the predictions made by the generative model. The neural connections traditionally known as feedback connections carry newly generated predictions about the upcoming sensory signal.

It should become clear now that the predictive-processing framework is complementary to the two general features of the brain addressed above: ongoing neural dynamics and massive feedback. The importance of massive feedback should be obvious. Similarly, a system that works by generating its own data is a system that one would expect to exhibit intrinsic ongoing dynamics. Recall that a common theme in the literature on ongoing dynamics was context-sensitivity. The predictive-processing framework requires context sensitivity in neural processing—the probability of perceiving particular objects changes with context. Overall, the predictive-processing framework looks to fit nicely within an understanding of the brain as an active, rather than a reactive, organ. With these general points in place, I will now turn away from a description of the general predictive-processing framework, and turn to evidence from the visual system in particular.[11]

The (once) standard feedforward model of visual neuroscience is heavily motivated by Hubel and Wiesel's discovery of receptive field properties for neurons in cat cortex (1959). Classical receptive field properties involve a neuron firing more strongly when a stimulus of a particular kind is presented in the receptive field of the neuron. Hubel and Wiesel, and subsequently many others, also found that neurons exhibit what is known as extra-classical receptive field properties (see Allman, Miezin, and McGuinness 1985 for a historical overview). An extra-classical receptive field response refers to the sudden decrease, or endstopping, in the firing of a neuron when the stimulus is extended to a region just outside of the receptive field of that neuron. There is no obvious way to explain these extra-classical responses within the standard model, but predictive processing can handle it in a straightforward manner.[12]

In a widely cited article, Rajesh Rao and Dana Ballard (1999) described a hierarchical neural network designed according to the principles of predictive processing described above. Higher levels in the network predict the responses at lower levels. If those predictions are not accurate, an error signal propagates to the higher levels through the bottom-up connections in order to correct future predictions. After Rao and Ballard trained the network on thousands of natural images, the simulated neurons in the network developed something similar to receptive field properties. Even more interesting, the network also developed extra-classical receptive field properties. For example, when a bar stimulus was presented in the receptive field of one of the simulated neurons, the neuron would fire strongly. But when a slightly longer bar was presented, so that the stimulus extends into adjacent areas, the neuron showed extra-classical endstopping. According to the predictive-processing approach, when a longer bar stimulus is presented, the simulated neuron would endstop because the higher-level prediction would have been more accurate. As Rao and Ballard explain, more accurate predictions for a longer bar should be expected, since the network was trained on natural images in which short isolated bars rarely appear. When the bar is longer, the higher-level simulated neurons are better able to predict the activity of lower-level neurons, and we see a decreased firing response on the lower level.

Rao and Ballard's work was at the beginning of a fruitful line of research using the predictive-processing framework to make sense of way the visual brain works. For instance, Tai Sing Lee and David Mumford (2003) have applied the framework to a wide array of evidence from visual neuroscience, including a consideration of more brain regions than Rao and Ballard. More recently, Michael Spratling has applied the predictive-processing framework to visual attention (2008) and to some of the details of neural response in area V1 (2012a, 2012b).

Although the focus in this section has been on the predictive-processing framework, the larger, and more important, point is that there are strong theoretical approaches to visual neuroscience that support premise (2), and that these approaches account for the properties of the brain, ongoing dynamics and massive feedback, mentioned in sections 6.1 and 6.2 above. Not all of these theoretical approaches fall under the banner of predictive processing, at least as I have described it here. Above I have already mentioned the pioneering work of Llinás and Freeman. There are other instances of theories in visual neuroscience that support (2) without adhering to the details of the framework described above. These include theories that give a large role to synchronous neural oscillations (Varela et al. 2001, Engel and Singer 2001), and to associative memory along with the default mode network (Bar 2007). Still more options will be described in the following chapter. In that chapter, I will cover the way in which evidence in support of the dual visual systems hypothesis can be understood in a way that supports premise (2).

6.4 Summary

I hope to have shown that a strong case can be made in support of premise (2) based on evidence from neuroscience. It is important to be clear, though, that the dominant, or textbook version, of visual neuroscience remains one that is opposed to (2), at least in broad terms. In this last section of the chapter I would like to summarize the neuroscientific case for (2) and make a final conceptual point against the standard feedforward model.

The neuroscientific case in support of (2) can be summarized with the following three points. First, brain activity involves ongoing intrinsic dynamics, which is what one would expect if (2) were correct. The ongoing activity suggests that the brain may be able to generate predictions spontaneously and in a context-sensitive manner. Second, brain wiring includes massive feedback connections, both cortical and subcortical. One would expect massive feedback as a straightforward way to implement the anticipation-fulfillment structure of (2). Third, there are a number of empirically motivated theoretical approaches to visual neuroscience, given in the previous section, which all share (2) as a common theme.

I would like to make a quick rhetorical point before moving on to the evidence for dual visual systems. It is commonly accepted in the philosophy and science of vision that the standard feedforward model jibes with visual phenomenology (Tye 2000; also the Jackendoff/Prinz argument from chapter 1). An understandable reaction to a new predictive alternative in visual neuroscience is to assume that the structure of the underlying neural processing is not reflected in visual phenomenology (Clark 2013b: 16), and then accept this disconnect as a limitation of the theory. The argument of the first part of this book reveals another option. Theories that fall under the general theme of (2) can appeal to my case for (1) in order to maintain that their empirical theories are phenomenologically plausible. Of course, phenomenological plausibility might not be on everyone's list of desiderata for a good theory in cognitive neuroscience, but I suppose it cannot hurt.

7 The Dorsal Stream and the Visual Horizon

7.1 Visual Consciousness and the Two Streams

There is a large body of evidence from cognitive neuroscience that supports a distinction between two cortical processing streams in the human visual system. The goal of this chapter is to suggest a way of interpreting the empirical evidence that is amenable to premise (2). The two cortical systems are the ventral stream, which projects from primary visual cortex to inferotemporal cortex, and the dorsal stream, which projects from primary visual cortex to posterior parietal cortex. The evidence regarding the two streams has been used in support of theories about visual consciousness. The most influential of these theories has been developed by David Milner and Melvyn Goodale (1995, 2005), who have focused on the functional output of the two streams. They have argued that the ventral stream is devoted to "vision for perception," which can contribute to conscious experience, and the dorsal stream is devoted to "vision for action," which cannot in principle contribute to consciousness (2010: 75). Their theory and most of its rivals are driven by empirical evidence, with little regard for the phenomenology of visual experience.[1] As a result of this methodology, there is currently no account of visual consciousness that accommodates both the empirical evidence about the two cortical streams and my claim (1) from the first part of the book. Not only is there no such account, but the existing theories appear to be in tension with vision understood as a process of anticipation and fulfillment. At least prima facie, understanding vision in terms of anticipation and fulfillment is at odds with Milner and Goodale's functional distinction between "vision for action" and "vision for perception."

The problem, as I see it, is as follows: there is a wealth of empirical evidence regarding the two streams and their roles in visual experience, but there is no theoretical framework for this evidence that is sensitive to the phenomenology of vision, to the view that I have been defending in this book, in particular.[2] In this chapter I will try to provide such a framework. The most relevant insight is, once again, from Husserl, who emphasized that visual phenomenology always includes a spatial and temporal fringe, or horizon. There is always an indeterminate periphery in space, and there is always anticipation of the next instant in time. These features of visual phenomenology are neglected in some of the most well-known literature on the two visual streams. But, as I intend to show, it is precisely these features that best explain the differences between the two streams.

I am going to review the evidence that *the crucial difference between the two cortical streams is in their spatiotemporal processing, rather than functional output: the dorsal stream processes peripheral retinal input with a high temporal resolution, and the ventral stream specializes in foveal input with less temporal resolution.* Following my suggestions in the previous two chapters, the general form of the processing in these streams is one of anticipation and fulfillment. Toward the end of this chapter, in section 7.6, I will discuss models that include anticipatory signals from the dorsal to the ventral stream. My suggestions about input differences can be found in the existing empirical literature, but there is yet no way to understand them in relation to conscious visual perception. The contribution of this chapter is to show how the phenomenological description of vision found in part I of the book can actually help us make sense of the disparate bits of empirical evidence. One way to express my main claim of this chapter, then, is as follows: dorsal stream processing plays a main role in the spatiotemporal limits of visual perception, to what Husserl identified as the visual horizon.

In addition to providing a new, phenomenologically motivated interpretation of the empirical evidence, my thesis in this chapter can clear up some areas of disagreement in the existing literature. For instance, Milner and Goodale (2010) are at odds with Yves Rossetti, Laure Pisella, and their colleagues over the way to describe the role of the dorsal stream (Pisella et al. 2006, Rosetti et al. 2010). The former maintain that dorsal processing is "vision for action" in the "here and now." The latter have emphasized that dorsal processing is devoted to peripheral vision. The account I develop offers a synthesis of elements from these two approaches.

There is also some disagreement about whether dorsal processing can make a contribution to conscious experience. Milner and Goodale have claimed that it cannot, in principle, make a contribution (2010: 75), and at least two prominent philosophers have relied on their claim in philosophical argumentation. Peter Carruthers uses this claim as "the grounds for one of the main arguments" against first-order theories of consciousness and in favor of a higher-order approach (2005: 201). The important part of the hypothesis, for Carruthers, is the output of the streams:

> Importantly for our purposes, the outputs of the dorsal system are unconscious, while those of the ventral system are phenomenally conscious (in humans). (2005: 200)

Similarly, in a critique of Alva Noë's sensorimotor approach to perception, Ned Block has appealed to the suggestion that dorsal stream output is not available for consciousness:

> [Noë's enactive view] would still clash with the facts about the two visual systems, since the enactive view would dictate that the (in fact unconscious) dorsal states are conscious. (2005: 270)

My task here is neither to evaluate higher-order theories of consciousness, nor to defend Noë's enactivism against Block's critique. Instead, I hope to show that there is a good alternative to the received understanding of the two visual systems, to the understanding of dorsal output as unconscious vision for action and ventral output as conscious vision for perception. As a part of my interpretation of the empirical evidence, I suggest that dorsal processing can contribute to conscious experience in the form of the visual horizon.

The neurophysiological division of cortical visual processing into two streams is widely accepted in the current literature (Gangopadhyay, Madary, and Spicer 2010), but there are important voices of dissent. Careful hodological work in nonhuman primates suggests that there are three, rather than two, cortical streams (Rizzolatti, Luppino, and Matelli 1998, Rozzi et al. 2006, Gallese 2007). In particular, the suggestion is that the dorsal stream should be further divided into a dorso-dorsal stream and a ventro-dorsal stream. Gallese proposes: "The dorso-dorsal stream has the characteristics suggested by Milner, Goodale and Jeannerod when they describe the dorsal stream as a whole" (2007: 3). Thus, the points I make about the spatiotemporal properties of dorsal processing can be taken as applicable to the dorso-dorsal stream. A discussion of the ventro-dorsal stream is left for further research.

In the next section of the chapter I will illustrate the phenomenological description of the visual horizon. In the third section I will outline the neurophysiological and anatomical evidence that input to the dorsal stream differs in spatiotemporal properties from input to the ventral stream. In the fourth section I will outline the relevant evidence from localized cortical damage and visual illusions. In the fifth section, I will cover cases of akinetopsia in order to explain how they can fit within the general framework of anticipation and fulfillment. In the final section I will mention how my main claim finds support in models that include anticipation in the neural dynamics of the visual system, models similar to those covered in the previous chapter in support of premise (2).

7.2 Introducing the Visual Horizon

Milner and Goodale famously suggested that dorsal processing is for action (and is unconscious) while ventral processing is for (conscious) perception. As noted above, some philosophers have uncritically embraced the dichotomy between vision and action (Block 2005, Carruthers 2005: 72). Other philosophical traditions, though, have long opposed such a dichotomy.[3] The close link between perception and self-generated movement was first discovered in the Western philosophical tradition by Aristotle (particularly in *On the Soul*). More recently, this link has been explored in detail for over a century now within the Husserlian phenomenological tradition (see the appendix for some details). Given the assumption that these philosophical traditions contain valuable insight regarding action and perception, we have a good motivation to resist Milner and Goodale's dichotomy. What I hope to demonstrate here is that there is a way to accommodate the empirical evidence without following Milner and Goodale in placing a wedge between "vision for action" and "vision for perception."

As discussed in chapter 1 and chapter 5, visual perception and action are intimately related—recall the "weak claim" from section 1.2. The process of anticipation and fulfillment typically involves anticipation of the sensory consequences of one's self-generated movements. The role of action in visual perception highlights the ever-changing indeterminate *spatial* fringe of our experience (more on this shortly). Similarly, the notion of anticipation incorporates a *temporal* fringe of visual perception: according to Husserl, all perception essentially involves the anticipation of the immediate future. To sum up: action and perception are closely linked through the cycle of anticipation and fulfillment, and this cycle always includes a spatial and temporal fringe. Following Husserl, we can refer to this spatial and temporal fringe as the *horizon*. I should be clear that my use of Husserl's terminology does not imply strict fidelity to his philosophical project. As Kristjan Laasik (2014) has shown, and I address in the appendix, there are important ways in which my appropriation of Husserl departs from what Husserl himself was doing.

Husserl uses "horizon" (*Horizont*) in a number of different ways in his corpus. Anthony Steinbock identifies three senses of the term: visual horizon, substantive horizon, and transcendental horizon (Steinbock 1995: 105–106). It is only the first sense of horizon, the visual horizon, that will concern us here. The visual horizon is spatial because of peripheral indeterminacy, and it is temporal because it involves experiences that are possible in the most immediate future. As Steinbock explains, "the horizon is not only conceived as a spatial halo, but also as a temporal court or fringe projected by the object" (1995: 105; also see Husserl 1977: 58). The visual horizon is a constant feature of visual experience, and it includes both a spatial as well as a temporal fringe.

The spatial aspect of the visual horizon has already been introduced through the discussion of peripheral indeterminacy from chapter 2. When we look at objects, there is always a spatial horizon, which occurs at the periphery of our visual field.[4] In principle, we can always move our eyes or bodies to gain a better perspective on that which we can only grasp indeterminately from the current perspective. This peripheral indeterminacy marks the spatial horizon of vision. For added illustration, one can return to Dennett's exercise for noticing the indeterminacy of the spatial horizon (1991: 53–54, and section 2.3 of this book). The human visual field has a small point of clarity (corresponding to the fovea on the retina), which is surrounded by a horizon of indeterminacy.

In addition to the spatial fringe, visual perception includes a temporal fringe of possible percepts. The important point here, again already covered in chapter 2, is that the most basic elements of visual perception always occur within a temporal structure. The implicit anticipation of how appearances will or could change lies at the temporal horizon of visual perception.

This horizon is the indeterminate border between present and future possible structured visual experiences. As time unfolds, the indeterminate anticipation of how a novel object will look from a hidden side can become determinately fulfilled as one moves around to view the object from the previously hidden side. Thus, Husserl describes the horizon as a "determinable indeterminacy" (Husserl 1966 §1).

The main point of this section of the chapter is that there is an indeterminate spatial and temporal horizon to vision. Dennett's example reveals the spatial indeterminacy, and the notion of visual anticipation shows temporal indeterminacy. In what follows, I shall take it as accepted that there is an indeterminate spatial and temporal horizon to vision.

7.3 Input to the Dorsal Stream

Now that the general phenomenological theme of the chapter—the visual horizon—has been introduced, here is the neurophysiological and anatomical justification for my main claim. The last few decades of work on the physiology of the primate visual system has shown that the input to the dorsal stream differs from the input to the ventral stream.[5] The most relevant differences for present purposes are that *the dominant input to the dorsal stream is processed faster and is less foveally concentrated than input to the ventral stream*. Recent results have made it clear that we should not oversimplify the differences between the inputs to the two streams, but all parties seem to agree that there are differences nonetheless. Here is a summary of the relevant evidence regarding these differences. Readers who are not interested in the neurophysiological details may wish to skip to the next section of the chapter.

There are two major types of parallel pathways from the retina to the thalamus and then on to cortex in primates.[6] The magnocellular pathway projects from the retina to layers 1 and 2 of the lateral geniculate nucleus and then terminates in layer 4Cα of the primary visual cortex. The parvocellular pathway projects from the retina to layers 3 through 6 of the lateral geniculate nucleus and then terminates in layer 4Cβ of the primary visual cortex (Kveraga 2007, Nassi and Callaway 2006).

In their classic paper, Livingstone and Hubel (1988) reported four key differences (speed, contrast, color, and acuity) in processing characteristics of the two pathways as discovered mostly through anatomical and

The Dorsal Stream and the Visual Horizon

physiological studies in nonhuman primates. The response of the magnocellular pathway is faster and more transient than that of the parvocellular pathway, and the magnocellular pathway is more sensitive to low-contrast stimuli. The parvocellular pathway is sensitive to changes in wavelength, unlike the magnocellular pathway, and is thus responsible for color processing. Also, the parvocellular pathway has smaller receptive field centers on the retina, and thereby has a higher acuity than the magnocellular pathway. To summarize, the magnocellular pathway is faster and more sensitive to contrast, and the parvocellular pathway processes color and with better acuity.

What does all of this have to do with the dorsal and the ventral streams? Livingstone and Hubel suggested that "the temporal visual areas [the ventral stream] may represent the continuation of the parvo system, and the parietal areas [the dorsal stream] the continuation of the magno pathway" (1988: 744). Subsequent research has revealed that this proposal is an oversimplification; the magnocellular and parvocellular do not map on to the dorsal and ventral streams in such a straightforward manner.[7] Importantly, though, Livingstone and Hubel's suggestion is not completely false, either. Milner and Goodale report that "most of the input to the dorsal stream is magno in origin" (1995: 36). More recently, Nassi and Callaway have filled in some details about the input to the dorsal stream. They focus on input to area MT (medial temporal), which "is primarily a processing station within the dorsal stream" (Milner and Goodale 1995: 50). Nassi and Callaway have found that input to MT "originates almost exclusively in [magnocellular] dominated layer $4C\alpha$" (2006: 12792). They point out that MT probably receives parvocellular input as well, but this input is likely more indirect and "may require additional synaptic relays" (ibid.). The main point to be taken from these details is the following: the magnocellular stream does not neatly map on to the dorsal stream, but it does constitute the dominant input to the dorsal stream. This finding is especially striking when one considers that the parvocellular stream "is tenfold more massive" than the magnocellular (Livingstone and Hubel 1988: 748).

Recall that magnocellular processing is fast and contrast-sensitive, yet color blind and has less acuity. It is likely that this is the nature of the information processing that dominates the dorsal stream as well. These results motivate my claim that dorsal processing differs in its *temporal* properties from ventral processing. There is neurophysiological evidence that dorsal processing differs in *spatial* properties as well, evidence that retinal input to the magnocellular pathway is distributed across the retina in a different manner than input to the parvocellular pathway. To be more precise, the parvocellular pathway is especially concentrated on foveal input, whereas the magnocellular pathway is not. Using intracellular staining techniques on intact human retinas isolated in vitro, Dacey and Petersen (1992) investigated the dendritic field size of the retinal cells that input to each pathway. Although the density of the cells that input to both pathways increases toward the fovea, they think it is likely the density of cells that input to the magnocellular pathway "increases more slowly approaching the central retina than does [density of cells that input to the parvocellular pathway]" (1992: 9669). Furthermore, they suggest the cells that input to the parvocellular pathway outnumber the magnocellular input cells by roughly 30:1 in the fovea. If their conclusions are correct,[8] the imbalance between magnocellular and parvocellular input is most extreme in the fovea. I should also mention here that there is evidence for cortical magnification of central vision in the ventral stream. Such magnification may be reduced in the dorsal stream (Milner and Goodale 2010, Colby et al. 1988, Brown, Halpert, and Goodale 2005)

To sum up these neurophysiological and anatomical details, the input to the beginning of the dorsal stream is not concentrated in the fovea, and this input is delivered to cortex faster than the input to the ventral stream, which receives high acuity input concentrated in the fovea. The differences in the processing based on retinal location continue in the cortex. Now I turn to the evidence from localized brain damage, which supports my claim that the differences between the two streams are mainly in the nature of their spatiotemporal processing.

7.4 Localized Damage and Illusions

In this section of the chapter, I focus on two important areas of evidence for the two visual streams. First I will discuss the visual and visuomotor deficiencies brought about by localized damaged to the cortex in humans, and then I will discuss evidence from actions directed toward illusory stimuli.

For my discussion of localized brain damage, I focus on the cases of patients D.F. and S.B., who both suffer from visual form agnosia caused by damage to the ventral stream. In addition, I will discuss recent work with patients suffering from optic ataxia, a condition caused by damage to the dorsal stream. These cases of localized damage add more support for my suggestion that the dorsal stream is involved with processing of the spatiotemporal limits of visual perception, or what Husserl would call the visual horizon.

The case of patient D.F. is well known from the work of Milner and Goodale over the last few decades. While in her early 30s, D.F. suffered bilateral damage to her inferotemporal cortex from carbon monoxide poisoning. This damage to the ventral stream has impaired her perceptual recognition abilities, but she is nonetheless able to perform, as normal, a variety of visually guided motor tasks. Milner and Goodale have appealed to evidence from D.F. to make the case that the ventral stream processes vision for perception and the dorsal stream processes vision for action. Their theory suggests that D.F. can still perform visually guided actions because of her intact dorsal stream.

Crucially, the kinds of actions that D.F. can perform are limited to actions of a particular temporal nature. As Milner and Goodale put it, D.F. is only able to perform actions in the "here and now" (1995: 137). Unlike in a normal subject, when there is as little as a 2-second delay between her view of an object and the initiation of her grasp of that object, D.F.'s grip size does not correlate with the width of the object (Goodale, Jakobson, and Keillor 1994). Milner and Goodale conclude that the dorsal stream processes vision for action, but only for "real-time" practiced actions. Motivated by the other lines of evidence herein, I would reverse the emphasis in the explanation. I have emphasized that the dorsal stream is marked by its particular dynamics, by processing what comes next visually. These dynamics are at work when we perform "real-time" practiced actions because it is such actions that push the temporal limits of visual perception. In other words, rather than say that the dorsal stream is concerned with actions, but only fast actions, I would say that the dorsal stream operates at a particular time scale, and that this time scale is especially useful when we grasp in a natural, practiced manner.[9] This way of describing the function of the dorsal stream moves away from the simple dichotomy between perception and action that shows up, for example, in the passages from Block and Carruthers cited above (section 7.1).

One might ask the following question at this point: If the dorsal stream plays a role in our perception of the visual horizon, and D.F. has an intact dorsal stream, then why does D.F. not consciously perceive the visual horizon? Perhaps she does consciously perceive the visual horizon, at least as much as anyone consciously perceives it. Note that D.F. does have conscious visual experiences of color and texture (Milner and Goodale 1995: 125–126). As Morgan Wallhagen has suggested, it is possible that D.F. also experiences other visual features, but that she is unable to conceptualize and thereby report on those features (2007: 556). An excellent candidate for the kinds of perceptual content that would be unconceptualized—by both normal subjects and visual form agnosics—would be the spatial and temporal fringe of the visual horizon. I do not have the concepts to describe the indeterminate content of, for instance, Dennett's playing card in the visual periphery, or the sign on the side of train passing the platform at full speed. Here is one place where Husserl is especially helpful: With care, we are able to describe some of the *structure* of visual phenomenology even though we might not pin down all the *content* in a satisfying manner (I return to the importance of distinguishing content from structure below in section 8.1.). The structure of the visual horizon could very well be preserved in D.F. without her reporting any particular perceptual content because the visual horizon in normal subjects does not lead to reports of particular content. Admittedly, the case of D.F. alone is not sufficient to conclude that the dorsal stream makes, or can make, some contribution to phenomenal consciousness. In order to investigate this issue further, consider the case of S.B.

Perhaps the case of visual form agnosic S.B. is not as well known as D.F., but it is no less fascinating.[10] Sandra Lê and colleagues have presented the case as follows (Lê et al. 2002). At the age of 3, a case of meningoencephalitis left S.B. with cortical damage more extensive than that of D.F. After the illness, S.B. lost both of his ventral streams as well as his left dorsal stream. D.F., in contrast, retained intact dorsal streams and the ventral damage was not total (Milner and Goodale 1995). S.B. represents a case of vision with only one dorsal stream. Another important difference between S.B. and D.F. is the age at which their damage occurred. Because S.B. was so much younger at the time of the damage, he likely had greater cortical plasticity as an advantage in recovery.

S.B. experiences no colors, and shows the expected range of deficits of ventral stream damage, including the inability to recognize objects and faces. The relevant question here is whether S.B. has conscious visual experience. The evidence indicates clearly that he must. Surprisingly, he is able to "drive a motorcycle and ... easily catch two table tennis balls at the same time and juggle with them" (Lê et al. 2002: 59). Also, he "is bothered by high luminance levels; he prefers to move within a low luminance world (dawn, night)" (Lê et al. 2002: 71). He has no problem moving about in an unfamiliar environment (ibid.). And this is all with only one dorsal stream.

At the very least, the case of S.B. shows that dorsal stream processing *can* contribute to conscious experience. Now, the early plasticity of S.B.'s cortex after the damage might mean that S.B.'s dorsal stream is connected in ways that are not to be found in normally developed subjects. Thus, it would be wrong to conclude that the dorsal contribution in normal subjects is precisely everything that S.B. experiences. The important point, though, is as follows. If and when the dorsal stream makes a contribution to conscious experience, we should not expect it would be a contribution that could be easily described. Therefore, the fact that D.F. cannot report on properties that are probably processed by her intact dorsal stream does not entail the conclusion that she has no visual experience of those properties. She could experience them, but in the way that we experience the temporal and spatial horizon, not unlike the way in which S.B. visually experiences the world.

The final set of cases to mention here are cases of optic ataxia due to dorsal stream damage. Milner and Goodale have emphasized that D.F.'s intact dorsal stream can enable her to perform visually guided actions in the "here and now." The temporal constraints on the nature of actions enabled by dorsal processing motivates my redescription of dorsal processing as having to do with the temporal limits of visual perception. I am also suggesting that the two cortical streams differ in the spatial nature of their processing. This claim finds support in recent articles by Yves Rossetti, Laure Pisella, and their collaborators. Based largely on their work with optic ataxics, they have argued that the dorsal stream processes peripheral information, while the ventral stream focuses on central vision. In particular, optic ataxics tend to perform normally, or nearly normally, on actions directed to objects in central vision (Rossetti, Pisella, and Vighetto 2003, Pisella et al. 2006). This finding is consistent with what one might expect from the non-foveal nature of the magnocellular dominant input to the dorsal stream (section 7.3).

In response to this line of reasoning, Milner and Goodale (2010) have emphasized that there have been optic ataxics who have shown deficits grasping in central as well as in peripheral vision (see Goodale et al. 1994). In line with my thesis about the two streams, though, one can explain deficits in natural grasping by appealing to the temporal properties of dorsal processing. Natural grasping is a fast motion that would require the high temporal resolution of dorsal processing.

As Rossetti et al. indicate, patients with optic ataxia do not complain of general deficits in vision for action. Instead, they tend to be aware of their disability when they are unable to perform skillful actions quickly; for instance, they complain of "a slowness and clumsiness in writing" (Rossetti et al. 2003: 177). Also, optic ataxics sometimes report difficulty with the exploration of a new and complex environment, such as a busy train station (ibid.). This difficulty is what one might expect with a deficiency in peripheral vision, because such a deficiency could alter the pattern of saccades one would normally make in exploring a novel and rapidly changing environment. To restate things in terms of the horizon, the visual horizon is especially valuable when exploring novel and dynamic environments. In such situations, we need to anticipate visually the way things will change, and we need to detect unanticipated changes in the periphery. With damage to the dorsal stream, and, I suggest, a subsequently compromised visual horizon, optic ataxics have difficulty coping with such situations.

Another main source of evidence for suggesting a functional dichotomy between the two cortical streams comes from experiments using visual illusions. Perhaps the most well known of these experiments involve the Titchener circles (figure 7.1). Salvatore Aglioti and colleagues (Aglioti, DeSouza, and Goodale 1995) showed that normal subjects fall victim to the illusion perceptually, but that their grip aperture reflects the true (nonillusory) size of the circles. Milner and Goodale take this result as further evidence for a functional dichotomy between vision for action and vision for perception. The question of whether action falls victim to perceptual illusion has been pursued widely in the last couple of decades. Here is not the place to review the sizable literature, but I would like to make two quick points, which, I think, are fairly important to keep in mind.

The first point addresses the central/peripheral distinction. If there is a difference between ventral and dorsal processing that reflects central versus peripheral vision (Pisella et al. 2006), then it will be important to

The Dorsal Stream and the Visual Horizon 143

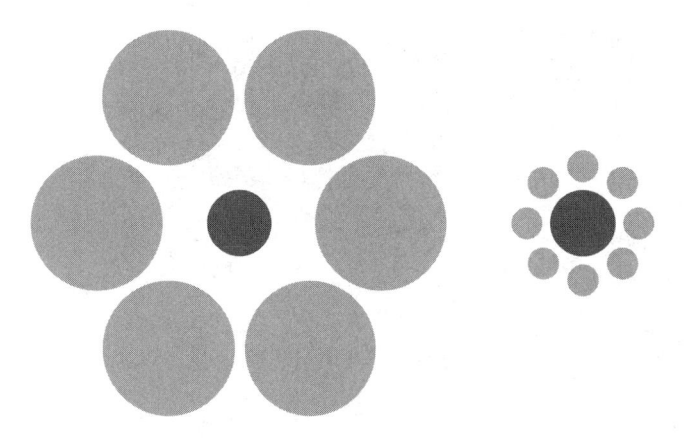

Figure 7.1
The Titchener circles.

consider how the strength of illusions vary with position in the visual field. For instance, the peripheral drift illusion (made popular with Akiyoshi Kitaoka's "rotating snakes"; see figure 7.2) occurs only in the periphery. In contrast, when viewing the Titchener circles, subjects presumably saccade between the two sets of circles in order to bring each set into central vision. Indeed, the Titchener circles illusion is so subtle that it is not clear whether we *can* experience it beyond central vision; that is, by fixating on a point that places both sets of circles in the periphery. Thus it may not be sufficiently precise to claim simply that an illusion fools conscious vision, since conscious vision can be central or peripheral.

The second point involves the dynamics of the experiments. The illusions occur when subjects are allowed to view the stimulus in an unhurried manner, but the grasp, which purportedly is not victim to the illusion, occurs quickly. Subjects are instructed to reach naturally, which de facto means to reach with some speed. Indeed, there is evidence that slowing the movement brings on the illusion (Rossetti et al. 2005, Króliczak et al. 2006). In addition, subjects who are instructed to reach in an awkward manner fall victim to the illusion. After subjects practice the awkward grip, the illusion no longer affects their reaching (Gonzalez, Ganel, and Goodale 2006, Gonzalez et al. 2008). Assuming that increased skill means increased speed, these results further support my claim that it is fast vision, not vision for action, which is supported by dorsal processing.

Figure 7.2
"Rotating snakes" (used with kind permission from Akiyoshi Kitaoka).

So if—and this remains a matter of debate—there is a dissociation between experiencing the illusory stimulus, on one hand, and visually guided grasping of it, on the other, the dissociation is not between conscious vision for perception and unconscious vision for action. Rather, the dissociation is between judgments based on central vision, on one hand, and fast grasping movements, on the other. The illusions in question do not occur in fast, visually guided actions, nor is there evidence that they occur in peripheral vision. Both of these abilities, fast movements and peripheral vision, are supported by dorsal processing, or so the evidence indicates. Thus, one could conclude that the dorsal stream is not fooled, as it were, by the illusion, and still maintain that the dorsal stream is devoted to the limits of spatial and temporal vision, to the visual horizon.

In this section I have tried to outline evidence from localized damage as well as the perception of illusions in normal subjects. I have argued that the key differences between the streams are spatiotemporal, rather than differences between action and perception. In addition to visual agnosia, another form of visual disturbance that might, prima facie, seem to challenge premise (2) can be found in various forms of akinetopsia. I turn to that topic in the following section.

7.5 Disturbances of Visual Motion

In this section I will digress slightly from the topic of the two-visual systems in order to address various kinds of disturbance of visual motion, including akinetopsia, or "motion blindness." I do so here because akinetopsia is similar to visual form agnosia in that is a well-known visual disturbance in the cognitive neuroscience of vision, and one that, from what I can tell, has not yet been explored from the perspective of anticipation and fulfillment. Initially, one might regard cases of akinetopsia to pose a challenge for the understanding of visual experience in terms of anticipation and fulfillment. If one sees moving objects as a series of static images, then the smooth, ongoing process of anticipation of fulfillment seems to be disrupted rather dramatically. Akinetopsia, then, could be presented as a counterexample to AF in the following way: individuals with akinetopsia no longer experience a continuous process of anticipation and fulfillment, but they are still able visually to perceive the world. It might seem, then, that anticipation and fulfillment are not necessary structural features of visual perception. More specifically, the second constraint from chapter 2, the temporality constraint, seems to be called into question. There I claimed that all visual experience occurs in time. But in cases of akinetopsia, patients have reported a visual experience as of a series of static visual images. Here I wish to show that akinetopsia provides no real challenge to AF. Quite the opposite, AF offers a nice way of starting to make sense of the disturbances reported in akinetopsia.

Before entering into the way in which akinetopsia can be understood in relation to AF, it is important to define some terms. There are a range of disturbances having to do with the visual perception of motion, and the terminology to describe these phenomena has not always been used in a fixed manner.[11] Perhaps the most common form of disturbance of visual motion is what is known as "trailing," in which the trajectory of an object in motion includes static images of previous locations of that object. Users of LSD and other hallucinogens commonly report experiences of trailing while under the influence of the drug and sometimes afterwards, as in cases of hallucinogen-persisting perception disorder (DSM-IV code 292.89). Trailing is not a real form of motion blindness because subjects remain able to perceive the moving object's true spatiotemporal location. Trailing is similar to a visual disturbance known as palinopsia, in which subjects experience afterimages of objects after those objects have been removed. The main difference is that trailing involves discrete images of moving objects, whereas palinopsia is a more general term for afterimages (with various levels of determinacy) of objects. (One could regard trailing as distinct from palinopsia, as in Dubois and VanRullen [2011], or as a form of palinopsia, as in Gersztenkorn and Lee [2015].) Palinopsia has a number of etiologies, including hallucinogens and hallucinogen-persisting perception disorder as well as brain damage (from lesions or head trauma), seizures, migraines, and nonhallucinogenic drugs (Abert and Ilsen 2010, Gersztenkorn and Lee 2015).

The most extreme, and rare, disturbance of visual motion is akinetopsia (motion blindness), in which subjects perceive moving objects as a series of static images. Unlike trailing and palinopsia, individuals with akinetopsia are not able to perceive the accurate spatiotemporal location of objects in motion. The most well-known case of akinetopsia is patient L.M. (Zihl, von Cramon, and Mai 1983), who suffered bilateral damage to area V5/MT in 1978 (Zihl et al. 1983, Shipp et al. 1994), though other cases have been reported (Zeki 1991, Pelak and Hoyt 2005):

[L.M. complained of] a loss of movement vision in all three dimensions. She had difficulty, for example, in pouring tea or coffee into a cup because the fluid appeared to be frozen, like a glacier. In addition, she could not stop pouring at the right time since she was unable to perceive the movement in the cup (or a pot) when the fluid rose. ... She could not cross the street because of her inability to judge the speed of a car, but she could identify the car itself without difficulty. "When I'm looking at the car first, it seems far away. But then, when I want to cross the road, suddenly the car is very near." She gradually learned to "estimate" the distance of moving vehicles by means of the sound becoming louder. (Zihl et al. 1983)

When tested, L.M. was unable to see the continuous motion of stimuli moving faster than 10°/sec. Importantly, her other visual abilities appeared to be unaffected, abilities such as visual acuity, stereopsis, and color vision (ibid., Hess, Baker, and Zihl 1989).

Now consider trailing, palinopsia, and akinetopsia as motivations for a possible objection to vision as anticipation and fulfillment, to premise (2) of the Main Argument, in particular. An initial, and understandable, reaction to these kinds of disturbances is to claim that the visual system operates with a series of discrete "snapshots" of the world. In disturbances of visual motion, those snapshots fail to be properly integrated into a seamless stream of consciousness (see Robson [2014] for a presentation of such a view for a popular audience). This move might also be presented as a challenge to my temporality constraint from section 2.2.

Apart from independent problems with the snapshot conception of visual experience (Hardin 1988: 7–18, Noë 2004: chapter 2), I suggest that there are two additional reasons to reject the snapshot interpretation of disturbances of visual motion. First, this interpretation generates the huge puzzle of how all of these snapshots are seamlessly integrated during normal vision (see my remark on trans-saccadic perception in section 1.3 above). In other words, if, against (2), temporality is not already an inherent property of visual processing, then the experience of the visual world in time requires a further ad hoc explanation. It may be less troublesome and more elegant, I suggest, to consider disturbances of visual motion to be deviations from the normal, temporally extended operations of visual perception rather than to regard the nonpathological temporal flow as an additional component of the normal snapshot-taking visual system.

The second reason to reject the snapshot explanation of disturbance of visual motion is that such explanation neglects self-generated motion. All of the disturbances of visual motion described above—trailing, palinopsia, and akinetopsia—are disturbances of the visual perception of objects that are in motion relative to the subject. If akinetopsia revealed the true snapshot nature of visual experience, then we should expect reports of snapshot experiences for self-generated movement as well. I have found no such reports in the literature. One case of a visual impairment for self-generated motion gave rise to complaints of vertigo and nausea, not snapshot vision.[12] Akinetopsia may be an important phenomenon for understanding vision, but, when carefully considered, it does not support a snapshot conception of visual experience.

Now, consider how disturbances of visual motion might fit within the general framework of anticipation and fulfillment, beginning with trailers and palinopsia. In order for the ongoing process of anticipation and fulfillment to work smoothly, perceptual fulfillments must not linger in visual consciousness. They must fade into unconsciousness in order to allow for future sensory fulfillments. More precisely, they must be *retained* in consciousness (see the discussion of retention in section 4.1), but not in phenomenal consciousness, not in the form of visual sensations. The straightforward way to understand trailers and palinopsia is to say that they are cases of sensory fulfillments that linger in phenomenal consciousness after they should have faded into a nonsensory retention. In the case of trailers, it seems that the fact that the object is in motion has something to do with the cause of the lingering fulfillment. In other cases of palinopsia in which the object is not moving, the sensory fulfillments of views on stationary objects linger in consciousness. On my view, trailers and palinopsia are better understood as a failure to discount or fade out previous fulfillments rather than a failure to "stitch together" distinct visual snapshots.

Akinetopsia represents a more drastic disturbance of the visual perception of motion than do trailers and palinopsia. Recall that, in akinetopsia, subjects are unable to perceive a continuous trajectory for objects moving faster than 10°/sec. Note that it is not a complete breakdown of the cycle of anticipation and fulfillment. The cycle remains normal for normal self-generated movement as well as for slowly moving objects. When the cycle does break down, in cases of objects moving with some speed, subjects only perceive a series of static images of those objects. Those static images can be treated in the same way that I treated trailers and palinopsia above, as lingering fulfillments. The difference with akinetopsia, though, is that in addition to lingering fulfillments, subjects also experience gaps in the fulfillments of their anticipations about the future trajectory of the object. There are two possible explanations here, in accordance with premise (2). The first one would treat akinetopsia as a disorder of fulfillments, and the second would treat it as a disorder in generating anticipations. First, one could say that akinetopsia shows cases of missing, or gappy, fulfillments. Normal subjects experience continuous fulfillments for objects in motion, whereas subjects with akinetopsia experience intermittent fulfillments, which manifest subjectively as a series of static images. The second explanation would treat akinetopsia as a deficit in generating anticipations. That is, when objects move with some speed, perhaps subjects with akinetopsia are unable to generate anticipations quickly enough. On this second option, there is no possibility of fulfillment due to the fact that there are no anticipations that can be fulfilled. In principle, it may be possible to investigate these two options experimentally. If patients with akinetopsia show some residual ability to predict the trajectory of quickly moving objects despite their compromised experience of those objects, perhaps in a forced-choice paradigm, then such an ability would tend to support the first option.[13]

As a way of bringing the discussion back to the main theme of this chapter, I would like to end this section by mentioning experiments on akinetopsia and its relation to the two visual systems. Thomas Schenk and colleagues (2000) tested akinetopsic patient L.M.'s ability to reach for moving objects. They were interested in determining whether motion blindness has an effect on, in Milner and Goodale's terms, "vision for action." They found that L.M. could reach for objects moving at 0.5 m/sec or less, but that successful reaching required relatively longer observation times as well as visual feedback from her moving hand. They conclude, based on these results, that area MT/V5 provides input to both the action and perception streams.

If one were to follow Milner and Goodale's functional account of the two streams, one might say that it is obvious akinetopsia affects vision for perception, and that Schenk et al. demonstrated that it also affects vision for action. But this interpretation is in tension with the passages cited above in section 7.3, indicating that area MT/V5 receives mostly magnocellular input and that it is, in Milner and Goodale's own terms, "primarily a processing station within the dorsal stream" (1995: 50). If area MT/V5 is primarily in the dorsal stream, and the dorsal stream is concerned with unconscious vision for action, then we should expect akinetopsics such as L.M. with MT/V5 lesions to show a primary deficit in vision for action rather than vision for perception.[14] Instead, patient L.M. shows a deficit *both* in perceiving quickly moving objects and in reaching for quickly moving objects. The view that I am defending avoids this tension nicely. Since I have been emphasizing the spatiotemporal differences in the cycle of anticipation and fulfillment for the parvo/ventral stream and the magno/dorsal stream, L.M.'s deficit in both perceiving and reaching for visual stimuli of a particular (fast) spatiotemporal scale fits nicely with my view. In the final section of this chapter, I will discuss some recent neurocomputational models of vision that give the dorsal stream a role in visual anticipation on a faster spatiotemporal scale.

7.6 Computational Models of Dorsal Anticipation

Up until now I have only outlined the evidence that dorsal processing is faster and less foveally concentrated than ventral processing. Now I would like to connect this claim with premise (2), and to suggest that the visual horizon is characterized by visual anticipation. In this section of the chapter I will mention some recent models of visual anticipation that include the dorsal stream.[15]

The predictive-processing models of neural dynamics from the previous chapter suggest that anticipation is distributed and widespread throughout the massive feedback connections in cortex and thalamus. What I have emphasized here, though, is that visual processing occurs on two distinct time scales, with the magnocellular stream delivering input to cortex slightly faster than the parvocellular stream. Given this fact, it seems reasonable that the faster processing scale might play a special role in providing anticipatory feedback for the slower, but more accurate, processing in the ventral stream. Indeed, several variations on this idea have been developed.

Combining evidence from visual masking and neurophysiology, Haluk Öğmen and Bruno Breitmeyer have developed a retinocortical dynamics (RECOD) model in which visual perception always involves a succession of three temporal phases: feedforward dominant, feedback dominant, and reset (Öğmen 1993, Öğmen, Breitmeyer, and Bedell 2006). The time differences in the magno- and parvocellular streams play a key role in determining temporal phases in visual perception. Although the focus of this model is subcortical, the model could be compatible with the suggestion that anticipatory feedback occurs in cortical dorsal processing as well.

Moshe Bar has developed a similar model, but one that gives a more central role to cortical processing (Bar 2003, Kveraga, Ghuman, and Bar 2007). In particular, his model proposes that the magnocellular stream delivers a fast sketch of the visual input to the orbitofrontal cortex. From there, feedback is sent to ventral processing area IT in order to facilitate object recognition. Bar's model leaves it open whether the fast input to orbitofrontal cortex is delivered via the dorsal stream or subcortical pathways. In any case, the dorsal stream could be understood here as a part of a larger system, a system characterized by the nature of its temporal and spatial processing:

The network employed in top-down facilitation of object perception may be part of an older system that evolved to *quickly* detect environmental stimuli with emotional significance. This primarily may involve *scanning the environment* for threat and danger cues, but also could include the detection of other survival-related stimuli, mating or food-related cues. (Kveraga et al. 2007: 160, emphasis added)

This passage lends further support for my claim that it is important to consider spatial and temporal dynamics when comparing the two cortical processing streams. It also adds the further fascinating suggestion that affective states may play an important role in anticipatory visual processing (also see Barrett and Bar 2009).

The final model of anticipatory neural dynamics is not terribly unlike Bar's, and, of the three here, it fits best with my main claim that the dorsal stream plays a key role in the visual horizon. In a series of articles, Jean Bullier (Nowak and Bullier 1997, Bullier 2001a, 2001b) has made the case that the magnocellular stream, as well as areas in parietal and frontal cortex, *including the dorsal stream*, constitute what he calls the "fast brain" system. This system provides feedback to earlier areas V1 and V2 in order to facilitate processing of information delivered by the parvocellular stream. Such a functional role for the dorsal stream in the "fast brain" could naturally be understood in terms of the temporal horizon and cited in support of premise (2).

One final point to mention here is that these models generate a novel empirical prediction, making for another example of "front-loaded phenomenology" (Gallagher 2003). If, as suggested, the dorsal stream plays some role in visual anticipation, then dorsal stream damage should bring about a decrease in object recognition speed. Evidence of such a decrease would be further evidence in support of my claim about the two streams.

The currently dominant understanding of the neurophysiological distinction between visual processing streams in cortex is based on the distinction between action and perception. What I have tried to show here is that there is another option, an option that goes far in accommodating the empirical evidence and supports premise (2). This other option is inspired by philosophical work that maintains a close link between action and visual perception. This alternative, based on Husserl's concept of the visual horizon, is that the difference between the two streams is chiefly a difference in spatiotemporal processing—that is, the dorsal stream deals with fast processing of peripheral information and the ventral stream deals with slower processing of foveal information. Also, as others have suggested, the dorsal stream may play a special role in anticipatory visual activity, which adds more detailed support for (2).

Part III

8 The Convergence

Now that I have made the case for both premises of the Main Argument, I will turn to some general methodological issues. This chapter is brief, and consists of two parts that address possible objections to the Main Argument, along with one final section that addresses the question of perceptual representations. In the first part, I discuss the Main Argument itself and defend it against a possible objection. In the second part of the chapter, I approach methodological questions having to do with ways of studying the mind and the brain. In particular, I suggest that symbolic dynamics is a methodology in cognitive neuroscience that is flexible enough to accommodate the range of claims that I have argued for in the first two sections of the book. In the final part of the chapter, I discuss the way in which my view might offer a middle ground between the defenders and deniers of internal mental representations. Let us begin by returning to the Main Argument in its entirety.

8.1 Back to the Main Argument

Recall the Main Argument of the book:

(1) *The descriptive premise*: The phenomenology of vision is best described as an ongoing process of anticipation and fulfillment.
(2) *The empirical premise*: There are strong empirical reasons to model vision using the general form of anticipation and fulfillment.
(AF) *Conclusion*: Visual perception is an ongoing process of anticipation and fulfillment.

The first part of the book is intended to support the first premise, and the second part the second premise. Now consider the argument as a whole.

Let us assume, throughout this chapter, that both premises of the Main Argument are true. In this section I will address an objection to the Main Argument that is motivated by the historical tension between subjective and empirical methods of investigating the mind. My reply to this objection highlights one of the virtues of the Main Argument; namely, that it offers some insight into overcoming this historical tension.

One slightly odd feature of the Main Argument is that AF is supported by the first premise and the second premise independently. If one true premise states that the structure of visual perception is one of anticipation and fulfillment, then AF is true. Why should we need two premises to do the job of one? We need two premises because of problems inherent to the study of the mind. On one hand, some have a strong intuition that we are able to discover truths about our minds through first-person reflection. On the other hand, natural science has an impressive track record for discovering truths about the natural world, including minds, and these truths often conflict with introspection (Schwitzgebel 2008). The history of the sciences of the mind is filled with tension between these two methods of investigation. Recall, for example, the opposition of behaviorism to unobservable mental states in the early and middle twentieth century, or the difficulties famously raised by Thomas Nagel (1974) for objective investigations into consciousness.[1]

Taking a lesson from that history, one might raise an objection to the Main Argument by claiming that these two methods of investigating a faculty of mind do not in fact have the same explanandum. To be pluralistic (or dualistic) about it, one might say that there is the subjective mind, which has properties to be investigated from the first-person perspective, and there is the objective mind, which has properties to be investigated empirically. Or, along with some researchers in the field, one might simply hold that the best—and only—way to study the conscious mind is to do so empirically. Dennett, for example, is known for advancing this sort of view about consciousness (1991).[2] These considerations can be taken to motivate an objection to the Main Argument; the objection would be that history shows us that one ought not to combine results in an unqualified manner, as AF does, from different ways of investigating the mind. History shows us that, at best, these different methods do not even have the same explanandum.

The motivation for that objection is the history of conflict between first-person and empirical methods of studying the mind. I would like to respond to that objection by pitching the Main Argument as a kind of conflict resolution, for it suggests that both of these methods of investigation can converge, *do converge*, on the same results for the case of vision. One of the virtues of the Main Argument is that it reveals a strategy for how one might reconcile phenomenology and the mind sciences. In this way, this book is an attempt to follow others who have sought to open the dialogue between phenomenology and the mind sciences,[3] and, more generally, an attempt to foster a dialogue between the humanities and the sciences (Snow 1963, Brockman 1995). Following in this tradition, I suggest that the Main Argument offers some insight for deepening this dialogue, for avoiding conflict between subjective experience and empirical results.

In this regard, perhaps the most important feature of the Main Argument is that it focuses on the structure, rather than the content, of visual experience. This strategy has roots in seminal work from consciousness studies. We have already seen a version of it in the Jackendoff/Prinz argument from the first chapter. David Chalmers cites Jackendoff, along with Moritz Schlick (1938) and Thomas Nagel (1974), as expressing support for his "principle of structural coherence" (1996: 224–225). This principle suggests that there is a systematic structural correspondence between phenomenology, on one hand, and psychology and neuroscience, on the other. Chalmers quotes a line from Nagel that puts the point nicely: "Structural features of perception might be more accessible to objective description, even though something would be left out" (1974: 449). The idea common among all of these thinkers is that progress may be more likely in relating conscious experience to the natural world if we seek that relation through the structure, rather than the content, of conscious experience.[4] The Main Argument is an example of that idea put to work for human vision.

The most dramatic cases of purported conflict between subjective experience and empirical science involve questions about the content of subjective experience, cases, for example, in which experiments reveal subjects to fabricate their own reasons for acting (Nisbett and Wilson 1977, Carruthers 2009). In one well-known experiment, subjects were told that they were participating in a "consumer survey" and asked to choose the best quality pair of stockings out of four pairs that were in fact identical. Subjects showed a strong right-hand bias in their selections, but gave reasons for preferring one pair over the others based on perceived quality (Nisbett and Wilson 1977: 243). Even if, based on results such as these, we grant that there is a conflict with regard to content, this conflict does not carry any implications about a conflict with regard to structure. As Nagel indicates, limiting ourselves to structure might leave something out, but the limited scope of a particular methodology is no reason to abandon or doubt that methodology. Also, even if Nagel is right that the focus on structure leaves something out, it still may be the case that a proper account of structure can offer insight into the general nature of content. In the following chapter, I explore this possibility further.

So far in this section I have made the case for AF and offered a response to what I take to be the most powerful general objection to the Main Argument. The convergence shows one way that there can be harmony between the subjective and the objective ways of investigating the mind. As noted in part II of the book, it is not always clear how this harmony is supposed to work. In particular, the subjective term used to describe the structure of vision is "anticipation," whereas the term commonly found in psychology and neuroscience is "prediction." While I do not think that a fine-grained analysis of possible semantic differences between these terms will be worthwhile, I should offer a comment.

Ultimately, I leave it an open empirical question as to the precise relation between the consciously accessible structure of anticipation and fulfillment, on one hand, and predictive processing, on the other. It is important to be clear that I am not suggesting there is a one-to-one mapping between visual anticipations and predictive neural signals. In the first part of the book, I suggested that visual anticipations cannot be individuated, so hunting for an isomorphism is not recommended. The important point, once again, is that the general structure is the same. Both (personal-level) anticipations and (subpersonal-level) predictions have content about the immediately upcoming sensory experience.[5] Both are future-directed. I suggest that this sameness of structure is sufficient for the convergence of (1) and (2) in the form of AF.

Due to the fact that the predictive-processing framework is relatively new to the philosophical scene, there is not yet much discussion of its relation to personal-level content in the literature. In a brief remark on the topic, Jakob Hohwy has suggested that we should understand predictive processing in terms of competing hypotheses in the brain, and that "[t]he hypothesis that is selected determines perceptual content" (2013: 37). He goes on to suggest that "perceptual content *is* the predictions of the currently best hypothesis about the world" (2013: 48). When applied to vision in particular, I take these claims, especially the second claim, to support the Main Argument, though with two important qualifications. First, I would like to add that the predictions include sensorimotor counterfactuals about how things would appear if the perceiver were to move in particular ways (Seth 2014). The scope of these counterfactuals is discussed above in section 3.2. On my view, some of the predictions are empty counterfactuals, and some are fulfilled as we move. I would like to resist Hohwy's suggestion that "We are not consciously engaging in … the way sensory input is predicted and then attenuated" (2013: 201). The first part of this book is intended to describe how the anticipation (and fulfillment) of sensory input is precisely the format of visual consciousness.

The second qualification has to do with Hohwy's description of perceptual content as being determined by *the* selected hypothesis. Although I am not certain whether Hohwy intends it, his way of describing content makes things sound more neat and tidy than I think they usually are. Perceptual content as hypothesis selection works well for some of the cases that Hohwy is considering, such as binocular rivalry. But in unconstrained natural viewing conditions, it may not be the best way to describe perceptual content and its relationship to subpersonal Bayesian processing. While we explore, we gain more evidence in support of the factual content with regard to one object, and we lose perceptual evidence for the factual content with regard to other objects as they fade out of view. The pattern of exploration and the content of the visual anticipations all depend on our interests and goals. This process is continuously unfolding in time—unlike the discrete selection of global hypotheses. As I argued in chapter 4, there is no determinate proposition that captures the content for any particular instant of visual experience. Instead of one hypothesis fixing conscious content, I suggest that visual content is a more direct reflection of the dynamic and probabilistic (messy) nature of the underlying processing (Madary 2012b).

The similarity in structure between visual experience and subpersonal visual processing brings the discussion into a new and challenging theme in the predictive-processing literature, the distinction between "surprise" and "surprisal" (Tribus 1961, Friston, Thornton, and Clark 2012, Clark 2013b section 4.1). "Surprise" refers to the personal-level experience of surprise, while "surprisal" refers to error in subpersonal predictions. Both the relationship between anticipation and prediction and the relationship between surprise and surprisal should be empirically tractable, and in similar ways. For example, mismatch negativity refers to the neural response after a violation of perceptual anticipations, and can be detected using EEG (Näätänen, Gaillard, and Mäntysalo 1978). The neural basis of mismatch negativity, as well as of surprise more generally, are ongoing research themes within the predictive-processing framework (Garrido et al. 2009, Garrido, Dolan, and Sahani 2011). It is precisely this sort of work that can uncover further details about the convergence between (1) and (2).

In this section I hope to have defended the Main Argument against the objection from the historical tension between subjective and objective methods of investigating the mind. My defense is that the Main Argument shows us a way of reconciling this tension in the case of vision. I have also made two subsequent points. First, reconciling the tension depends on focusing on the structure, rather than the content, of visual experience. Second, the details of the convergence between personal-level anticipations and subpersonal-level predictions should be empirically tractable, as evidenced by recent work in the predictive-processing literature.

8.2 The Best of Both Worlds—Symbolic Dynamics

In this section I would like to make a quick note about how the Main Argument can fit in with more general theoretical approaches within cognitive science. My purpose in covering this territory is to preempt an objection that my support for the Main Argument contains elements that are incompatible with each other. For instance, on one hand, my argument in chapter 3 in support of premise (1) required thesis (F), which is the claim that we represent factual properties in perception. The representation of factual properties is a persistent mental state that is most naturally modeled in terms of propositional attitudes, most at home in classical cognitive science. On the other hand, the process of anticipation and fulfillment, as well as ongoing neural dynamics, are marked by continuous change and are thus most naturally modeled using dynamicist cognitive science, which is often anti-representational (van Gelder 1995, Bechtel 1998). In order to avoid the objection, then, there must be some way of doing cognitive science that is friendly both to classical cognitive science as well as dynamicist cognitive science. In this section I will quickly present one version of such an approach, known as symbolic dynamics, and I will make some comments about how symbolic dynamics might be applied in the case of vision.

The framework that I am urging emphasizes the anticipation-fulfillment structure of the phenomenology of vision, and it appeals to evidence from perceptual psychology as well as neuroscience. Such a framework calls for the flexibility to include persistent mental states, such as the perception of factual properties, as well as models of transient underlying neural dynamics. This kind of flexibility can be found in hybrid approaches to cognitive science, approaches that accommodate the dynamics of cognition by use of a trajectory through multidimensional state space, but which also make room for persistent mental states by partitioning that state space.

One well-known example of this sort of approach would be Peter Gärdenfors' geometrical models of mental representation (2000). A similar approach can be found in an area of mathematics known as symbolic dynamics. It is an area of mathematics that has been developed over the previous century (Hadamard 1898, Morse and Hedlund 1938, Lind and Marcus 1995) and put to use in a variety of ways in the natural sciences, especially in physics. Symbolic dynamics offers a way of studying dynamical systems. As many readers will know, a dynamical system in mathematics is a way of describing a trajectory through a multidimensional state space. It is widely used in the natural sciences, and has been used in cognitive science (Thelen and Smith 1994, Port and van Gelder 1995, van Gelder 1995, Chemero 2011) as well as neuroscience (Izhikevich 2007).

The basic idea behind symbolic dynamics is to partition the state space into regions and assign a symbol to each region. In this way, the trajectory through the state space can be redescribed, in a coarse-grained manner, as a string of symbols. By describing the behavior of the system in terms of symbol strings, we move into territory familiar from classical cognitive science. We also gain the ability to include static, or persistent, mental states in a system that is in continuous change.

A number of scientists have already begun to explore the use of symbolic dynamics in cognitive neuroscience (Dale and Spivey 2005, Atmanspacher and beim Graben 2007, Spivey 2007 [chapter 10], Yoshimi 2011, 2012a, 2012b). The attraction of symbolic dynamics is that it allows the flexibility of dynamical systems without giving up the ability to use more traditional techniques, such as propositional attitude ascriptions, when needed. There are elements of my account that are more naturally modeled with dynamical systems, and elements that are not. For instance, perceptual anticipation can be modeled using attractor basins from dynamical systems theory (Spivey 2007, Friston and Kiebel 2009). In addition, the context sensitivity of perceptual anticipations, discussed in chapter 3, could be explained in terms of transitions within the attractor landscape itself (Tsuda 2001), or by hysteresis, which is a well-researched feature of dynamical systems (Strogatz 1994). As mentioned above, though, claims about representing factual properties in perception, or thesis (F) from chapter 3, is not the sort of thing that is conducive to dynamical systems theory. Representing a factual property is a coarse-grained way of describing the ongoing process of anticipation and fulfillment. Thus, when we have the need to investigate such coarse-grained mental states, then we could partition the state space using symbolic dynamics. To give a toy example, my perception of a teacup from different perspectives would involve the ongoing process of anticipation and fulfillment. This process could be modeled using dynamical systems. But the mental state described by my report, "I perceive that there is a teacup before me," is a state that remains static throughout that process of anticipation and fulfillment. If we want to capture that static propositional attitude, then we can partition the state space, creating a region that would include the entire trajectory of anticipation and fulfillment as I look at the teacup from different perspectives.[6] Although I will not explore it further here, this technique may be relevant for addressing the Davidson/McDowell epistemological worry from chapter 4.

8.3 Do We Need Internal Representations?

There has been a great deal of debate over the existence of mental representations in the past several decades. On one hand, the central theoretical commitment of classical cognitive science is that cognition is a process of computation over internal representations (Fodor 1981, 1987; Markman and Dietrich 2000). On the other hand, the positing of mental representations has met with strong criticism from a number of different perspectives (Varela, Thompson, and Rosch 1991; van Gelder 1995; Port and van Gelder and Port; Dreyfus 2002; Chemero 2011; Hutto and Myin 2012). Although I will not enter into the vast literature on this topic, I would like to discuss how some of the central themes of this book urge a particular way of thinking about visual representations, and mental representations more generally. The general point is that, on the view I am taking, the content of visual representations is always highly organism-relative (or subject-relative). This feature of visual representations marks a departure from the way in which mental representations are typically thought about in the debate.

As sketched in the previous chapters (especially chapter 4), visual anticipations are representational in the sense that they have content; they have accuracy conditions. But this content is not simply world-directed. On this view, the purpose of vision is not, as Marr famously put it, "to know what is where by looking" (1982). Instead, the purpose of vision is to serve the practical needs of the particular organism, to inform the organism about whether the relevant bits of the visual world appear as they should relative to the interests, goals, and perceptual history of that particular organism. Recall that where we look depends on what we are trying to achieve (as discussed in section 5.3), and the level of determinacy of the visual anticipations for each saccade depends on our familiarity with the visual scene (as argued in section 3.3). In these ways, the content of visual processing is closely bound up with the details of the organism doing the perceiving.[7]

Now consider the well-worn debate over mental—in this case perceptual—representations. The classical cognitivist claims that perception involves the creation of internal representations that are supposed to correspond to various parts of the perceiver's environment. An anti-representationalist, perhaps of the "4E" variety (Rowlands 2010; Ward and Stapleton 2012), denies that there are any such internal states. The organism-relative conception of perceptual content shares and rejects elements of both sides of this debate. In agreement with the representationalist (and disagreement with the anti-representationalist), I have argued that anticipatory visual states do have accuracy conditions. And it seems to me that the natural physical basis for such anticipations is some kind of internal model, as in the predictive-processing approaches discussed above.[8] But I depart from the standard conception of perceptual representations as being simply about the world. Instead, I have urged that the content of visual anticipations also reflects the organism's relationship to the world. The strong organism-relativity of my conception of visual representation is, I take it, at odds with mainstream representationalist commitments and, at the same time, very much in the spirit of anti-representationalist thinking.[9] Thus, the view that I am recommending can be construed as a path for peace in the representation wars (Madary 2015a, Clark 2015b).

My goal in this chapter has been to offer some remarks about how the Main Argument might fit in with larger methodological issues in the study of consciousness and in cognitive neuroscience more generally. In the final chapter of the book, I will assume that AF is true and argue that it offers a new way of describing many aspects of social cognition in humans. In particular, I will argue for the claim that the content of visual perception has a strong social element.

9 Seeing Our World

In chapter 4 of the book, I addressed some issues having to do with the structure of visual content. In this final chapter, I would like to advance some claims about the content itself. In particular, I would like to defend the claim that *visual content has a strong social element*. Call it thesis VCS. According to VCS, we see a world that we share with other humans, engaged in particular cultural practices. We see *our* world rather than *the* world. Visual content includes social and rational norms. The mental operations that enable intelligent social interaction—often relegated to the inner and unconscious realm of cognition—can actually unfold in plain sight, as it were.

This chapter consists of three parts. First, I make the direct connection between AF and social content, and clarify my position. Second, I present a range of empirical results that support my claim. Third, I return to some themes from the first chapter, themes about the general architecture of the mind.

9.1 AF Content Is of a Shared Social World

In this section I will develop the argument for VCS, which follows from AF. First I should note my intellectual debts: the idea that perceptual content is social is not a new one. It can be found in Husserl's posthumously published writings on intersubjectivity (consisting of three massive volumes, 1973a) as well as in his well-known late work, *The Crisis of the European Sciences* (1976 §54b). Similarly, Merleau-Ponty emphasizes that the social element is ever-present in experience:

> We must therefore rediscover, after the natural world, the social world, not as an object or sum of objects, but as a permanent field or dimension of existence: I may well turn away from it, but not cease to be situated relatively to it. Our relationship to the social is, like our relationship to the world, deeper than any express perception or any judgment. ... We must return to the social with which we are in contact by the mere fact of existing. (1945/1962: 362)

The idea can also be found in contemporary work within the phenomenological tradition (Zahavi 2005: chapter 6, Gallagher 2005: chapter 9, Thompson 2007: chapter 13).

What is new here, I hope, is that I am offering some details about *how* visual content is social. It is straightforward to talk about seeing colors and shapes, but how we see social content is not obvious. Although my own position differs slightly from hers, we can take Edith Stein's work on this topic (1913/1989) as a starting point. In her doctoral dissertation, supervised by Husserl, she applied Husserl's anticipation-fulfillment structure of perception to the way in which we empathize with others. She explains:

> The averted and interior sides of a spatial thing are co-given with its seen sides. In short, the whole thing is "seen." But, as we have already said, this givenness of the one side implies tendencies to advance to new givenness. ... The co-seeing of foreign fields of sensation [of other subjects—MM] also implies tendencies, but their primordial fulfillment is in principle excluded here. ... Empathic representation is the only fulfillment possible here. (1913/1989: 57).

Stein's suggestion, as I understand it, is that empathic awareness of the other is structurally similar to awareness of physical objects. The appearance of a table from a particular perspective serves as a fulfillment for my representation that there is a full table before me. The fulfillment involved in our perception of physical objects is the fulfillment of sensation, and is, in Stein's terminology, "primordial." The fulfillment involved in our empathic awareness of the other human being occurs through our perception of the other's animated body. Empathic fulfillment is not "primordial," but it does share the general structural similarity with perceptual fulfillment.

More recently, Joel Smith (2010) has offered an account of the way in which the mental states of others can show up in visual experience. His account also borrows from Husserl's notion of perceptual anticipation, and it turns out to be quite similar to Stein's results in some places, although he does not cite Stein as an influence. The way in which we perceive the mental states of others, Smith argues, is analogous to the way in which we perceive the hidden sides of objects through unfulfilled anticipations. He writes:

Just as the rear aspect of the book is visually present without being visually presented, so another's misery is visually present even though only their frown is visually presented. (2010: 739)

He goes on to defend this idea against some objections by introducing a way to individuate perceptual states according to functional role.

VCS shares similarities with the views that Stein and Smith have developed. All three of us rely on Husserl's notion of perceptual anticipation in order to identify the social content in perception. But my view is stronger than theirs in the following way. I am claiming that the social content in perception is pervasive. On my view, the social element of visual content goes beyond the content involved with attributing mental states to others. The sociality of visual content is primary in development, and it remains as an element of visual experience throughout maturity. The social aspect of perception is at play when we see each other, as Stein and Smith have argued, but it is also at play when we see our environment more generally—our homes and workplaces, our streets and sidewalks, our tools and food, and even the natural world.

If we accept that AF is true, then my claims about the social content of vision follow naturally. Visual anticipations are directed toward how the world might appear if we were to view it from different perspectives. Recall, once again, that visual anticipations have various degrees of determinacy. In more *familiar* environments, we will have more determinate anticipations. Similarly, the content of visual anticipations will be partly determined by previous experience, by how the world has appeared in the past. The influence of previous experience on AF content suggests that violations of anticipations would signal a departure from the *normal* course of appearances. Thus, familiarity and normality enter into perceptual content. Now, for creatures such as us, the familiar and the normal is social as a matter of fact. If familiarity and normality enter into the content of visual anticipation, and if the familiar and the normal are social for us, then visual content is social.[1]

Here is the argument for VCS step by step:

1. (AF) Visual perception is an ongoing process of anticipation and fulfillment.

2. Visual anticipations have various degrees of determinacy.
3. The familiarity and normality of the visual scene is reflected in the degree of determinacy of anticipations and the degree of fulfillment of anticipations, respectively.
4. Familiarity and normality are social properties for humans.

Conclusion: (VCS) Visual content has a strong social element (for humans).

The first premise is the main thesis of the book. The second premise was first defended in chapter 2 above. The third premise was defended immediately above, prior to my presentation of the Main Argument step by step. Perhaps I should add some comments about my fourth claim, that familiarity and normality are social for creatures like us. I only mean to make the uncontroversial point that the human being is a social animal. When we interact with others, there are norms of behavior. Violations of those norms appear to us in the form of violations of perceptual anticipations. Adherence to these norms appear to us in the form of fulfillments of perceptual anticipations. When we use tools or move about our homes, workplaces, and towns, doing so in a normal manner fulfills perceptual anticipations—both our own and for others who might see us. Doing so abnormally violates perceptual anticipations. It is in these considerations that one can see how my view goes beyond that of Stein and Smith. I am suggesting that the social content of vision goes beyond the perception of other humans. My view is that there is a social element in the perception of other humans *as well as* in artifacts and our environment more generally. It should now be clear what I intend when I claim to illustrate how visual content is social. The key point is this: *social factors partly determine the content of visual anticipations*. The violation of visual anticipations can signal the violation of social norms. Thus, social norms show up for us at the level of visual content.

There are three objections that I ought to address. First, one might ask about moments when we are alone, away from others, perhaps in pristine nature. In such situations, surely, one might say, there is no social content to what we see. I maintain that visual content remains social in such situations. My thinking behind this claim can be motivated by considering Merleau-Ponty once again:

Just as nature finds its way into the core of my personal life and becomes inextricably linked with it, so behavior patterns settle into that nature, being deposited in the form of a cultural world. (1945/1962: 347)

Even when one is alone, one still faces the choice of behaving in a way that would be socially normal or not. Behaving abnormally would bring about an unusual pattern of visual experiences that would, plausibly, alter the content of visual anticipations. When acting abnormally we may experience violations of anticipations, or a decrease in the degree of determinacy of anticipations. We can choose to act normally or abnormally, and perhaps no one will ever know what we choose, but the fact would remain that our visual experience would reflect whether we are acting in accordance with social norms, in accordance with the "deposited cultural world," or in violation of social norms.

The passage continues:

> Not only have I a physical world, not only do I live in the midst of earth, air and water, I have around me roads, plantations, villages, streets, churches, implements, a bell, a spoon, a pipe. Each of these objects is moulded to the human action which it serves. Each one spreads around it an atmosphere of humanity which may be determinate in a low degree, in the case of a few footmarks in the sand, or on the other hand highly determinate, if I go into every room from top to bottom of a house recently evacuated. (1945/1962: 347–48)

In other words, being alone does not remove the social nature of what we perceive. In the extreme case, when we are alone in untouched nature—not even footmarks in the sand—we have the unusual case of the lack of an "atmosphere of humanity." In such cases, we experience a kind of abnormality, the abnormality of having no sign of our conspecifics. Visual anticipations are disappointed because we normally anticipate traces of other humans in our perceptual environment. The disappointment of anticipations in such untouched environments may explain the thrill of being in such places.

The second objection is that humans are not necessarily social. We might consider a hermit or a feral child or even an unpleasant thought experiment in which a human child is artificially hatched and then nourished and raised by some impersonal machine. This objection reveals a limitation in the scope of my claim about visual content being social. It is not a universal claim about all human beings. My claim above is that visual content is social because visual content includes the normal and the familiar, and the normal and familiar are social. There may be cases, however rare, in which the normal and familiar are not social. In such cases, visual content may not be social.

Despite the fact that my claim is not a universal claim about all humans, it is still a claim that departs from standard ways of thinking about visual content. As covered in the first chapter, Jesse Prinz argues that visual content is merely of the surfaces and shapes around us (2012: 52). Some cognitive neuroscientists might go further and include object recognition in the content of visual experience. None of these standard approaches include the possibility of social content in vision.

The third objection is as follows. One might insist that we must first see nonsocial visual properties in order to then make inferences about social content. Along this line of thought, one might conclude that nonsocial content is primary and that social content is somehow secondary or even nonperceptual. After all, we first must see an object as a hammer—see its shape and surfaces—before there is any concern about the normal or familiar way in which hammers are used. My reply to this objection is to return to the fact that perception is always perspectival. One consequence of this fact is that perception of properties is always incomplete. It is, strictly speaking, not correct to claim that we see a hammer *simpliciter*. Seeing a hammer in a normal orientation, or being used in a normal manner, stirs up relatively determinate anticipations, which give us a sense of the familiar. Seeing the hammer in an unusual orientation stirs up less determinate anticipations, reflecting the unfamiliarity of the perceptual scene. Once we reflect on the fact that there is no simple, nonperspectival view of an object, this objection loses its force.

Now that I have offered my general descriptive motivation for social content in vision, I will turn to some of the empirical results that support my view.

9.2 Empirical Support

My primary argument for VCS is based on description of first-person experience, except for the uncontroversial observation that humans are social animals. But much of the debate over the nature of social cognition is interdisciplinary, with appeals to results from neuroscience and developmental psychology. In this section I wish to show that my position, which emphasizes the way in which social cognition can occur within visual perception, is supported by a broad range of empirical evidence.

Let us begin with some observations about human development. It is well known that newborns exhibit preferential viewing for faces (Johnson et al. 1991). This strong preference suggests that visual content is primarily social from the very beginning of our lives outside of the womb.[2] Developmental psychology also supports my claim that the social content of vision is not limited to instances in which we are looking at other humans. In an experiment that illustrates this point, Krista Casler and colleagues (2009) taught 2- and 3-year-old children to use novel tools. Later, the children observed a puppet using those same tools. When the puppet used the tools in an atypical manner, that is, differently from the way in which the children were taught to use the tools, the children protested. For instance, they told the experimenter how the puppet was using the tool incorrectly, or they tried to intervene in order to teach the puppet the correct use for the tool. These results suggest that children naturally apply a kind of normativity to the use of artifacts. They presuppose that there is the proper way in which one ought to use a tool.

Children's normativity regarding tool use is a natural fit with VCS and AF. Perceiving an artifact stirs up anticipations about the ways in which the artifact will appear when one uses or handles that artifact in a normal manner. The content of these anticipations is largely determined by social factors, by the way in which we have seen others, and perhaps ourselves, use the artifact. The protest of the children is triggered by the violations of their anticipations about how the tools should appear when they are being used. Now, I should acknowledge that the combination of VCS and AF is not the only framework that can be used to explain these results. A more classical approach could explain the results in terms of the beliefs and desires of the children. While the classical alternative is possible, I suggest that the appeal to VCS and AF is preferable. My reason for this claim is explanatory parsimony. On my view, we have an explanation of visual perception that meets all three of the constraints from chapter 2 in addition to an explanation of a variety of abilities within social cognition. On the classical alternative, we must posit separate ad hoc mechanisms for perception and social cognition. Parsimony and simplicity favor my approach over the classical alternative. I'll return to this theme below.

If VCS is correct, one might expect correlation between differences in social cognition, on one hand, and differences in visual experience, on the other. There are at least two ways in which this expectation is met: in disorders of social cognition, and in cross-cultural comparisons. One of the main characteristics of autism is a disability in social cognition. If vision has a strong social element, then we should expect differences in visual experience for individuals with autism. The evidence supports this expectation. Many open questions remain as to the details, but it is becoming clear that there are differences in, for instance, facial processing (Behrmann, Thomas, and Humphreys 2006) and motion perception (Kaiser and Shiffrar 2009) for individuals with autism.

The second area in which we see a correlation between differences in social cognition and differences in visual experience is in cross-cultural comparisons. If vision has a strong social element, then we might expect that one's culture can have an impact on how one sees the world. Anecdotally, we might expect this impact to reveal itself in patterns of saccades as determined by common social interests. Experimental evidence also shows such an influence of culture on vision. Studies have shown that East Asians and Westerners tend to differ in their "cognitive styles," with East Asians being more holistic and Westerners being more analytic (Masuda and Nisbett 2001). Recent work on this theme suggests that the differences occur in the allocation of resources for visual attention (Boduroglu, Shah, Nisbett 2009), which offers further support for the tight connection between vision and social content. Our cultural background, it seems, partly determines our visual attention.

Another well-documented source of evidence for the social element in visual content is something that readers can see for themselves. For over 30 years now, psychologists have known that some dynamic point-light displays are naturally perceived by us as walking human beings (Johansson 1973). By fixing lights to various parts of a walking human, such as the hands, feet, hips, and head, psychologists create stimuli that are completely dark except for the little lights moving in a particular way. The surprising feature of these displays is that we cannot help but perceive the moving lights as a human walker. I urge readers to experience such displays for themselves.[3] Not only do we see the lights as humans, but we see more particular properties of the walker, including the sex (Kozlowski and Cutting 1977) and the emotional state (Dittrich et al. 1996).

Consider how the point-light walkers can be described in terms of anticipation and fulfillment. We naturally spend a good bit of our visual lives watching other humans walking around. We need to do so in order to avoid collisions on busy sidewalks or in a stroll across campus. Thus, the motion of the point lights traces a pattern with which we are familiar. When we first see the moving lights, anticipations are stirred up based on previous (and common) experiences of others. As the lights move in a familiar pattern, those anticipations are fulfilled and we experience the stimulus as a walking human rather than a two-dimensional display of strangely moving lights.

There is a great deal of further evidence that can be cited as lending support for VCS. In the remainder of this section, I will mention two related areas without entering into the details. These areas are social signals and cultural evolution. If VCS is true, then we would expect humans to use visual perception in order to receive, and eye movements in order to send, social information. There is a wealth of evidence indicating that we do so. In a survey article on this topic, Chris Frith (2008) covers the detection of mood from facial expressions and posture, gaze-following as a way of understanding intentions, and the chameleon effect (in which two interlocutors experience mutual respect and trust when they imitate each other's mannerisms). Building on some of the same lines of evidence, Michael Tomasello (1999) has argued that *cultural evolution* explains how humans have developed advanced cognitive abilities in the relatively short period of 6 million years of evolutionary history that separates modern human from other great apes. Cultural evolution is driven by the human capacity to collaborate through shared intentionality. Crucially, shared intentionality is achieved in large part through the ongoing cycle of action and visual perception. For instance, infants learn one aspect of shared intentionality around one year of age by coordinating their gaze "triadically" with their human interlocutor *and* with the object to which both of them are directed (Tomasello et al. 2005: 682).

My primary argument for VCS from the previous section was not based on empirical science. The point of this section has been to gesture toward the great deal of empirical evidence supporting VCS, evidence that reveals the strong social element of visual perception.[4] In the final section, below, I will draw some general lessons about mental architecture that are motivated by VCS.

9.3 Embedded Rationality

If I am correct about VCS, then a general lesson about mental architecture follows. Recall the discussion of mental architecture from the first chapter of the book. Borrowing from Susan Hurley (1998), we distinguished two competing ways of understanding mental architecture. The classical "sandwich," according to Hurley, treats perception as input and action as output. All of the heavy lifting is done by cognition, which is the real substance of the sandwich. Perception and action are just slices of bread holding it all together. Cognition, on this view, is often understood in terms of propositional attitudes, such as beliefs and desires, implemented in some kind of Fodorian language of thought. Hurley's alternative to the classical sandwich is a view in which perception and action are not mere input and output. Instead, they are interdependent in the form of an ongoing cycle on different temporal and spatial scales. So which is it, the sandwich or the cycle?

Each model has strengths and weaknesses. As one might expect, the strength of the sandwich lies in modeling abstract cognitive tasks, and its weakness—according to its critics—lies in modeling tasks that involve ongoing skillful behavior, behavior that requires fine-grained coordination between action and perception. The cycle has the exactly opposite profile. It excels in explaining ongoing skillful behavior, but has a weak spot when it comes to cognition or rationality. Hurley was aware of this shortcoming and offered an alternative conception of rationality, cited previously in section 1.2 (1998: 409). On Hurley's alternative, the mind is made of content-specific sensorimotor loops (1998:21). Rationality is not localized in a belief/desire processing engine, but rather emerges out of this vast layered network of loops.

Hurley's suggestion can seem a bit mysterious, especially when contrasted with the straightforward simplicity of the alternative, in which rational thinking can be expressed nicely using natural language. A skeptic might respond to Hurley's view with the following question: How does higher-level rational content emerge out of a system of sensorimotor loops?

I think that VCS along with some of the themes discussed above bring us slightly closer to an answer for the skeptic—at least for the case of social cognition. The best response to the skeptic's question is to deny the presupposition that perception is devoid of higher-level content. If we deny this presupposition, then there is no mystery about how high-level content might emerge.

One good reply to the skeptical question is to say that social cognition can emerge out of sensorimotor loops because social content is already present in visual experience.

The results about normativity in tool use for infants, cited above, give us one way to illustrate Hurley's idea of rationality through sensorimotor loops. The children's perception of the objects becomes associated with various motor representations that involve using the object correctly. Seeing the object stirs up anticipations about how the object might be used. When it is used "incorrectly," those anticipations are disappointed.

A closely related general line of evidence supporting Hurley's architecture suggests that *imitation* plays a large role for human intelligent behavior. Normal human perception of others performing an action may bring about an impulse to perform that same action, to imitate (Heyes 2011, Belot, Crawford, and Heyes 2013). We act rationally by inhibiting these impulses, by using our "veto power" over the urge to act (Kühn, Haggard, and Brass 2009). In some cases of brain injury (Lhermitte, Pillon, and Serdaru 1986), patients lose the veto power and exhibit utilization behavior (using tools at inappropriate times) or imitation behavior (imitating conspecifics at inappropriate times). These considerations motivate an understanding of human action in which perception and action are tightly linked through sensorimotor loops. This way of describing action and perception could be understood as a neurally based approach to Gibsonian affordances. Seeing the artifact activates the motor routine to use the artifact; in my terms from above, seeing the artifact stirs up visual anticipations about how the artifact might appear when being used normally. Seeing the other's smile activates the motor routine to smile. Rationality is, in part, a matter of stopping—vetoing—these routines when appropriate. None of this, of course, is to suggest that veto power in a system is sufficient for rationality. The claim is that veto power in mature humans may be crucial for the way in which we behave rationally.

In the first chapter of book (section 1.2), I mentioned Dennett's intentional stance as an alternative to the physical stance. Now we are in a position to see an alternative to both the intentional and the physical stances; one could call it the *embodied stance*. We take the embodied stance when we understand and predict rational behavior by attending to the cycle of action and perception—both our own cycle and others. The 2- and 3-year-old children in Casler's experiments take the embodied stance when they *see* the tools being used incorrectly. We all take the embodied stance when we interact directly with others according to norms, as doing so fulfills our implicit anticipations. All of this is not to suggest that the intentional stance is obsolete or that there are no other strategies or "stances" deployed for mindreading. We might use the intentional stance in situations when we evaluate or predict rational behavior without the help of perception. Recall Dennett's own example of the intentional stance at use in predicting Mr. Gardner's behavior based on what Mrs. Gardner says into the telephone (1987: 26). Tellingly, that case is one in which we are making predictions about someone who is not perceptually present. In such cases, the embodied stance cannot be used, and we need another strategy. But there are many situations in which we interact socially with others in the flesh, situations that do not require the intentional stance. We only need to detect deviations from the norm by detecting disappointed visual anticipations.

VCS reveals how we can take the embodied stance, how rational behavior is embedded in the cycle of action and perception. The main contribution in that regard is that VCS offers some detail about how perceptual content can include the kinds of things that are often relegated to the domain of cognition. One main advantage of the cycle is parsimony. The cycle seeks to explain intelligent behavior by appeal to the ongoing dynamics and interplay between two obvious abilities: action and perception.[5] The sandwich, including the intentional stance, posits a hidden third ability that is supposed to be doing all the work: cognition. Another main advantage of the cycle is that it finds support in "veto" models of action from recent cognitive neuroscience.

The main thesis of the book, AF, makes a substantive claim about the nature of visual perception. If it is correct, and I hope to have made a strong case that it is, then there are implications in philosophy, psychology, and neuroscience. In this final chapter, I have started to explore the implications for social cognition and for mental architecture more generally. My hope is that these initial considerations might foster future research on vision and the human mind.

Appendix: Husserl's Phenomenology

The purpose of this appendix is to address issues having to do with my appropriation of some elements of Husserl's philosophical project. I have chosen to include relatively few references to Husserl's writings in the body of the book in order to avoid distracting readers who are not interested in such things. The appendix is for readers who are interested in such things. My goals here are three. First, I wish to offer some of the details about where some of the core ideas behind the book can be found in Husserl's main writings.[1] Second, I would like to address methodological questions about the degree to which my claims are compatible with Husserl's larger project. In doing so, I will engage with Kristjan Laasik's recent work on that topic (2014). Third, I would like to address some of the contemporary literature on applying Husserlian phenomenology to cognitive science, including Francisco Varela's "neurophenomenology."

A.1 Finding AF in Husserl

As indicated in the first chapter, thesis AF has its roots in Husserl's thought. In the first of his *Logical Investigations*, Husserl makes a distinction between intention and fulfillment (1900/1993 I §§9 and 10). He continued to make philosophical use of this distinction throughout his writings. In the context in which it was first introduced, Husserl uses the distinction to offer a theory of the meanings of signs and language. What is relevant here is that he later applied the distinction to his analysis of perception.

On the relationship between intention and fulfillment, Husserl suggests that it is something that we have all experienced and gives us examples. Intentions expressed by linguistic utterances can be fulfilled when objects named become present to us in perception (1900/1993 I §9; Simons 1995).

Other types of intentions that can be fulfilled include expectations, hopes, fears, and wishes (1900/1993 V§29, VI§10). The intentions at play in perception are slightly different than the types of intentions just mentioned. Husserl is clear that perceptual intentions are not necessarily expectations in the way that we expect particular states of affairs to obtain. Rather, the kind of intentions at work in perception can be compared to the way in which we implicitly intend the pattern of a rug to continue where the rug is hidden by a piece of furniture (1900/1993 VI §10). As I covered in the third chapter, perceptual intentions are indeterminate (Husserl 1966: §1), usually implicit (Husserl 1900/1993 VI §10), and it is difficult to say how they relate to propositional content (Husserl 1900/1993 VI §§4–5; Husserl 1973b).

Husserl described perception by making a crucial distinction between partial intentions and total intentions. Partial intentions are correlated with adumbrations of objects, with the way objects appear from particular perspectives. Total intentions are correlated with the entire object, or with what I referred in earlier chapters of the book as the factual properties of the object. Importantly, partial intentions are always *fused* together to form a total intention. The fusion suggests that partial intentions are nonindependent parts of the total intention; they are incapable of existing except as parts of a total intention. The fusion also means that the partial intentions cannot be separated from one another, but rather flow into one another without distinction (also see Yoshimi 2016, section 5.2.1). Here Husserl is applying theoretical machinery developed in his third *Logical Investigation* (1900/1993 III §§8 and 9). New partial intentions are continuously stirred up (*erregen*) as other partial intentions are fulfilled (1900/1993 VI §10). This process of stirring up partial intentions is later investigated in his genetic phenomenology under the theme of "laws of propagation of intentional awakening" (1966: 151 / 2001: 198). Perception is a process constituted by the continuous synthesis of partial intention and fulfillment (Husserl 1966 §§1–4; Husserl 1973b §32).

Recall the discussion of the perspectival nature of perception from the second chapter. This feature of perception was a main concern for Husserl. He illustrates it with the example of a red sphere (1900/1993 V §2). When looking at a sphere that is uniformly red, we can notice shadows and highlights—perspectival variations in color—across the surface, but we also see that the sphere is uniformly red. Now let us apply Husserl's framework of intention and fulfillment to the example of the red sphere. The variations in color—the highlights and the shadows—fulfill our partial intention of the way the red sphere appears from our perspective. This intention is partial for two reasons. First, it is partial because it always implies other partial intentions, which may or may not become fulfilled. The way the color variation appears as I look at the sphere stirs up partial intentions about the way the color variation will appear if I rotate the sphere a few degrees, for instance. The second reason why the intention is partial is because it is always a part of the total intention, which is correlated with factual properties of the object. The total intention of the uniform redness is never adequately fulfilled, because we can never see the sphere from all possible perspectives. Still, the fulfillment of the total intention can increase as more and more partial intentions are fulfilled (1900/1993 VI §§16 and 24).

The fulfilling acts of perception are somehow enabled by sensations, but the details of this relationship remain controversial (see Hopp [2008] for some of the complications). In the case of the red sphere, the sensations are various color sensations that play a necessary role in the fulfillment of partial intentions. Importantly, the color sensations are not themselves the objects of intentions (Husserl 1900/1993 V §14). Instead, the color sensations acquire their own intentional content that enables them to participate in the act of fulfillment.

The structure of intention and fulfillment allows Husserl to illustrate his claim, in *Ideas I*, that "*The same* color appears 'in' continuous manifolds of color *adumbrations*" (1977 §41).[2] Partial intentions are fulfilled by acts enabled by color sensations. The color sensations play a role in fulfilling a partial intention which is always a part of a total intention. The partial intention is correlated with the perspectival color appearance, and the total intention is correlated with the uniform color of the sphere. Thus, we see the same red in the variations because the variations correlate with fulfilled partial intentions, which are fused together with unfulfilled partial intentions to form the total intention whose objective correlate is the uniform redness. As our perspective changes, we encounter new adumbrations of the object, which leads to further fulfillments and the stirring up of new partial intentions. Note that this description is not ad hoc; for Husserl, perception *always* involves the continuous interplay of intention and fulfillment.

Also, I should be clear that Husserl does not seem committed to intentions being fulfilled by movement, or by implicit knowledge of sensorimotor contingencies. This fact distinguishes Husserl from sensorimotor theories of perception (O'Regan and Noë 2001; Noë 2004). Husserl often emphasizes how movement is crucial for perception, but partial intentions are not essentially connected to the sensorimotor domain. One could say that perception for Husserl is constituted by the ongoing activity of anticipatory partial intentions, but this activity is not *necessarily* linked to motor activity. For instance, Husserl often uses the example of anticipating the development of a melody (1900/1993 VI §10, 1969 §7), which presumably is not a kind of sensorimotor anticipation.

The similarities between AF and Husserl's framework should now be obvious. But I should quickly address a discrepancy between AF and Husserl's framework as I have presented it. I have sketched Husserl's framework of *intention* and fulfillment, but AF is a thesis about *anticipation* and fulfillment. When first developing the framework, Husserl was dealing with linguistic meaning, not perception. When applying the framework to perception, one can be more precise about the nature of the empty perceptual intentions: they are anticipatory. The idea that perceptual intentions are anticipatory can be found in his work from the first decade of the twentieth century (1969, 1973b). Continuing in his later work, his *Analyses of Passive Synthesis* from the 1920s, Husserl ties in perceptual intentions with his work on time consciousness (Husserl 1971) and refers to them as protentions (*Protentionen*) (Husserl 1966: 7). In the same work, he refers to perceptual protentions as anticipations (*Erwartungen*, 1966: 13, and *antizipiert*, 1966: 7), and it is this usage that I have adopted.[3]

A.2 Descriptive Psychology or Transcendental Phenomenology?

There is at least one central way in which my appropriation of Husserl's thought appears to deviate from his philosophical project. I have deliberately avoided language about constitution and constitutive analyses. Readers familiar with Husserl will know that this kind of language is central to his philosophy. Kristjan Laasik (2014) has appealed to this deviation—among other things (see chapter 7)—in order to claim that my synthesis of vision science with Husserl's phenomenology is not really Husserlian. In particular, Laasik raises the following critical points against my appropriation of Husserl:

1. Husserl was interested in giving a constitutive analysis of objectivity, and such an analysis proceeds in levels of complexity that are missing in my account.

There is the level of the constitution of the spatial object in terms of anticipations, then there is the level of the constitution of the material object, then there is the level of our practical engagement with the Lifeworld. (Laasik 2014: 431–432)

2. The spatial margins of the visual scene are not what Husserl meant by horizons. Such margins only describe contingent psychological features of human perception, and thus cannot serve in the constitutive role that Husserl intended. (Laasik 2014: 431)

Before turning to some of the details, I would like to make one important general point in response to the charge that my appropriation of Husserl's work deviates from Husserl's own philosophical commitments. The point is this: fidelity to Husserl's philosophical project is not one of my immediate goals. It is in the spirit of honest scholarship that I mention him and cite his passages. Thesis AF and some of the details in my defense of premise (1) are inspired partly by my reading of some of his works, and I ought not to hide that fact. Still, there are good reasons, I think, for exploring Laasik's criticism further. One thing to note about both of his critical points is that they stem from my avoidance of any kind of constitutive analysis. This issue, as it turns out, reflects a philosophically interesting tension, or a turn, within the development of Husserl's own thought. Let me explain.

Husserl's "breakthrough work," the *Logical Investigations*, is both long and ambitious, with the main goal being to rescue mathematics and logic from psychologism. This goal leads Husserl through central topics in philosophy of language (Investigations I and IV), metaphysics (II and III), philosophy of mind (V) and epistemology (VI). But for all of this scope and ambition, the *Logical Investigations* are less ambitious compared to Husserl's later transcendental phenomenology.[4] Sometime during the first decade of the twentieth century, between writing the *Logical Investigations* and the *Ideas I*, Husserl turned away from the realism of the former and toward the transcendental idealism of the latter. The motivation behind this turn is not obvious from his main works, but it has been explored a good bit in the secondary literature (see Ingarden 1975, for example). Like Kant before him, Husserl seems to have been led to transcendental philosophy through the problem of skepticism, especially through his reading of Richter's *Der Skeptizismus in der Philosophie*, the two volumes of which were published in 1904 and 1908 (Spiegelberg 1978 vol. 1: 134n1, Bernet, Kern, and Marbach 1993: 256n2). The path to constitutive analysis thus was opened up by Husserl's effort "to overcome global skepticism" (Zahavi 2008: 364). He becomes fully dedicated to this new conception of his project, emphasizing the transcendental aspect of phenomenology in his *Ideas I*. The transcendental project, for Husserl, involves giving a constitutive analysis of all objects of consciousness. Perhaps the best example of such an analysis is in his *Ideas II*, which was a major influence on Merleau-Ponty (Husserl 1989: xvi).

The important philosophical point here is that Husserl's notion of constitution reflects a larger philosophical commitment, a commitment to what would today be called a kind of anti-realism about the mind-independent world. Robert Sokolowski points out that Husserl's doctrine of constitution includes the assertion "that consciousness is a condition *sine qua non* for the real" (1970: 137). But Sokolowski adds:

There is nothing surprising about this assertion, nor is there anything idealistic about it. Real things, since they are units of sense, presuppose subjectivity as their necessary correlate, because sense can only arise in connection with intentional subjectivity. (ibid.)

Regardless of whether Husserl's transcendental philosophy is a form of idealism,[5] I think it is safe to say that the assertion above is one that would meet resistance from a good number of philosophers and scientists today, Sokolowski's casual defense of it notwithstanding. It is for this reason that I have avoided the aspects of Husserl's thought having to do with constitution.[6]

Recall the question that this book addresses: What is the general structure of visual experience? My strategy has been to address this question with as little philosophical baggage as possible. Thesis AF can be embraced by thinkers of all metaphysical stripes. One price that I pay for metaphysical neutrality is that I am open to the kind of criticism that Laasik raises: my project is not true to Husserl's at its most philosophical core. On the other hand, I do maintain that there is nothing about AF that is incompatible with the more ambitious project of transcendental phenomenology, just as I maintain that AF is not incompatible with various formulations of physicalism (Yoshimi 2015). One can, following Husserl, apply thesis AF in a constitutive analysis. Or one can apply AF when trying to make sense of the neural dynamics that, if you will, constitute phenomenal consciousness. But whether or not one puts AF to this kind of philosophical work has little to do with the truth of AF itself. Let us agree on AF despite any incompatible metaphysical proclivities that we may have.

Apart from constitutive analysis, Laasik's second critical point raises a question about whether the structure of visual perception as described by AF is a contingent psychological feature, or something perhaps philosophically more interesting. He notes that my descriptive support for AF suggests the latter (2014: 431). As I have presented it, I have made no commitments to strong modal claims about AF or about premise (1), the descriptive support for AF. As mentioned at the outset of chapter 2, by appeal to first-person description I have only meant to indicate features of visual experience that we can notice without using the methods of empirical science. That is, my support for (1) does not rely on experimental results. Similarly, premise 4 in my step-by-step argument for VCS in chapter 9—the premise that familiarity and normality are social properties for humans—is an observation about contingent features of our species, as one can see from my discussion of cases in which premise 4 does not obtain.

Going slightly beyond my stated position, we might consider whether there are good reasons to make claims about necessary features of perceptual experience. Husserl suggested that the perspectival nature of perceptual experience was a necessary feature; it is essentially true of every transcendent object that it shows up in (any) consciousness through perceptual adumbrations (1977 §44). Wayne Martin has noted that Husserl's view is open to a counterexample:

> Imagine some kind of conscious intelligence that is embodied in a kind of fog. We humans perceive an object from a single perspective, but the fogging consciousness simply fogs all around it, taking in all sides at once. Many animals manage to integrate sensory input from two sides of their body; why shouldn't the fogging being integrate views of an object from every side? (2005: 210)

In using this example, Martin was making a larger rhetorical point about the hopelessness of a particular style of argumentation. Still, though, his example has some force and is worth considering. I suggest that there is an important aspect of Husserl's insight that Martin misses here. Namely, the way that properties are given to us in adumbrations (perspectivally) is such that their appearances change in a way that is sensitive to self-generated movement—recall Siegel's doll thought experiment from chapter 3 for a contemporary take on this insight. The distinction between the subject and a transcendent object requires some kind of sensitivity to movement, or, in Siegel's terms, perspectival connectedness. In the case of the fogger, the appearance of the enfogged object ought to change in accordance with the movements of the fogger (or the movements of the object relative to the fogger). The object will appear differently to the fogger depending on its precise location within the fog, say whether it is partially immersed or totally immersed. Also imagine the fog turning itself relative to the stationary object. In that case, the fog would have access to changes in appearances relative to its own movements. Even sentient fogs, one might argue, should have perspectival connectedness to their objects.

Now, I have only suggested that Martin's initial counterexample to Husserl's necessity of perceptual adumbrations may fall short. It remains open to revamp the counterexample in order to meet my concerns, which is, I suppose, Martin's original point. He was warning about inevitable stalemates in counterexample battles. Whether or not Martin is correct with this larger rhetorical point, the particular case of the necessity of perspective for the perception of transcendent objects remains open. I do not insist on Husserl's claim that perspective is necessary for the perception of transcendent objects, but I do not know of any compelling reasons to reject it.[7]

A.3 Phenomenology and the Sciences of the Mind

So far I have discussed the way in which my main themes of this book is related to Husserl's transcendental phenomenology of constitutive analysis. This discussion has focused on critical remarks from Laasik. But there is also an existing body of literature in which results from the phenomenological tradition are applied to issues in the empirical sciences of the mind.[8] A seminal work in this field is Francisco Varela's (1996) proposal of "neurophenomenology" as a methodology for the scientific investigation of consciousness (also see Petitot et al. 1999). The "working hypothesis" of neurophenomenology is that "phenomenological accounts of the structure of experience in cognitive science relate to each other through reciprocal constraints" (1996: 343). Here Varela joins Chalmers (and others, see section 8.1 above) in drawing attention to the structure of experience as an important and worthwhile epistemic target. But he charges Chalmers' suggestion with being incomplete, because, for one reason, it "needs a method for exploration and validation." Varela's proposed method, neurophenomenology, involves giving a central role to trained phenomenological reflection within empirical studies, such as in subsequent work by Antoine Lutz and colleagues (Lutz 2002, Lutz et al. 2002, Lutz and Thompson 2003).

My own strategy in this book is very much inspired by Varela's proposal. One clear similarity is that I have followed Varela's recommendation that these issues be addressed in a strongly pragmatic spirit. Indeed, it is the pragmatic spirit that lies behind my deliberate metaphysical neutrality. But there are also important differences. First, unlike Varela, I do not see what I am doing as an attempt to solve the "hard problem" of consciousness. Second, and more concretely, I have not appealed to empirical research in which subjects are trained in phenomenological reflection. My approach, to return to themes from the first chapter, is a bit more like the one pursued by Jackendoff and Prinz, an approach in which we can identify a structural convergence between the phenomenological description of visual experience, on one hand, and empirically motivated models, on the other. That is, instead of conducting research with mutual constraints in mind, I have suggested that there is a natural convergence between two distinct ways of investigating visual experience. Of course, none of this is to rule out or disparage Varela's method; I only wish to be clear that I have not directly adopted it. Also, I do not mean to imply that the natural convergence, which is the theme of this book, is the end of the story. As I mentioned above in sections 7.6 and 8.1, there are specific areas in which my analysis raises questions for future empirical research. Overall, interdisciplinary research on visual consciousness has been working with a phenomenologically impoverished account of its structure, choosing to focus on its content instead. By being clear about the structure of visual experience, we can hope for more fruitful collaborative progress.

I would now like to address some critical questions raised by Tim Bayne (2004) directed toward the project of bringing Husserlian phenomenology to bear on issues in cognitive science. Bayne's first main criticism is that Husserlian phenomenology is neither unique nor privileged as a way of investigating first-person experience. He finds difficulty seeing how phenomenological reflection is any different from introspection. There is something to this point: What some people mean by the term "introspection" might capture, more or less, what Husserlian phenomenologists are trying to do. Varela himself (1996: 335) includes William James and the Kyoto School in Japan as practitioners of the kind of first-person investigation, or phenomenology, that he is recommending. The issue depends on what we mean by our terms.

If we follow Bayne himself on the meaning of these terms and define "introspection" as "an unmediated judgment that has as its intentional object a current psychological or phenomenal state of one's own" (Bayne 2015), then it is clear that Husserl is not interested in introspection as a proper topic of his phenomenology. He is not investigating "current" phenomenal states, but rather the abstract structural features—essential features, if you will—of all phenomenal states of different kinds. The interest of phenomenologists is closer to what Bayne defines as "indirect introspection," which "concerns the general nature of one's conscious experience and is not grounded in a single act of introspective attention" (2015: 2). He adds a parenthetical remark that indirect introspection "is not really a genuine form of introspection at all" (2015: 12). So, it appears that we have an answer to Bayne's question about what distinguishes Husserlian phenomenology from "mere" introspection: Since phenomenology is a kind of indirect introspection, it is not genuine introspection, at least on Bayne's terms. Now, Bayne's main focus is on cases of direct (genuine) introspection, while I have been trying to emphasize the value of indirect (Husserlian) introspection. I suggest that Bayne is wrong in his claim that "direct introspective judgments clearly have more warrant than indirect judgments" (2015: 12). The three constraints from chapter 2 offer examples of warranted indirect judgments. It should be clear, *pace* Bayne, that one's introspective judgment that, say, all visual perception is perspectival has more warrant than one's report of the content of one's visual experience at any particular instant. There are two general reasons to think that indirect introspective judgments can be more warranted than direct judgments. First, indirect judgments involve general structures over time, and thus do not face the difficulty of capturing the fleeting (and indeterminate) content of particular mental states. Second, indirect judgments are more amenable to intersubjective confirmation than are direct judgments. That is, it is reasonable to think—excessive skepticism notwithstanding—that it may be possible to investigate general features of experience in intersubjective discourse. The particular contents of one's mental states at any instant, in contrast, is, by its very nature, not intersubjectively evaluable.

Apart from Bayne's worries about how to situate Husserlian phenomenology in relation to other forms of introspection, I want to suggest that the proof of the value of Husserlian phenomenology can be found in its results. Any objection that does not seriously engage with those results is a red herring. As I have argued in chapter 4, describing the structure of visual experience as a process of anticipation and fulfillment is superior to the other views on offer in the literature. And that way of describing perception was first discovered and developed by Husserl. If one wants to investigate various aspects of first-person experience, one could do far worse than to overcome some contemporary prejudices and consider what Husserl had to say about it.

The reason behind the value of consulting Husserl's work is one that has been touched on above, and one that Bayne himself raises in his critical remarks; namely, Husserl's search for essential features of conscious experience, or essences.[9] In this search, Husserl turns away from the content of particular conscious states in order to find the general structures that always determine different kinds of mental states (to put it roughly). In Bayne's terms, he rejects direct introspection in favor of indirect introspection. This method, as I have argued in the case of perception, yields powerful results. Admittedly, and as discussed above, Husserl's own estimation of his method is somewhat high, claiming that these structures are essential for any form of consciousness. An important point, though, is that for the purposes of engaging in a fruitful exchange with the sciences of the mind, it does not matter whether these general structures are in place for conscious experience as such. After all, the empirical sciences of the mind cannot even begin to investigate merely hypothetical or supernatural conscious beings. Husserl's search for essences is at least valuable because of what it reveals about *human* conscious experience.

In bringing up this topic, Bayne worries that any insight into the essential structures of one's own consciousness may not generalize to other humans or to consciousness as such (2004: 353). His worry about consciousness as such is misplaced because of the reasons just given above. But what about generalizing to other humans? Here I think Bayne's skepticism is overextended. If one does not consider the details, it may be a reasonable point to say that one's own introspection may not generalize to other humans. But this point loses its force if we appreciate the abstract structural nature of Husserl's descriptions, and if we consider phenomenological reflection in the Husserlian tradition as a communal enterprise to be conducted over generations (Steinbock 1995: 258; 2003: 317–318). Consider, again, the case of perception. The three constraints sketched in chapter 2 are inspired by Husserl's treatment of the topic and they are, as far as I can tell, generalizable to all human visual experience. Similarly, Husserl's work on other elements of consciousness holds the promise of being generalizable to all human experience, work on topics such as time consciousness (Gallagher 2003, Varela 1999, van Gelder 1999), imagination (Husserl 1980, Marbach 1984), intersubjectivity (Husserl 1973a, Stein 1913/1989, Zahavi 2001), attention (Husserl 2004, Depraz 2004, Breyer 2011), emotion (Husserl 1966, Drummond 2009), embodiment (Husserl 1952, 1973b, Gallagher 2005), and so on. The proof of the value of Husserl's method can be found in the things he had to say on these topics. We must consider the detailed claims and investigate whether or not they indicate generalizable features of conscious experience.

Before moving on, I want to comment on another of Bayne's concerns about the usefulness of Husserlian phenomenology for cognitive science. In discussing the formal methods recommended by Varela and others to bridge phenomenology and neuroscience, he has suggested that "the explanatory itch would remain" (2004: 361). I agree, which is why I have made no promises about solving the "hard problem" or bridging the explanatory gap. But I have indicated that the convergence of results between phenomenology and empirical science may offer the opportunity for bridging connections using formal models (Yoshimi 2016); symbolic dynamics (in section 8.2); or category theory (Petitot 2000; discussed in section 4.4), for instance. My contribution here is not to propose detailed formal models, not to build the bridges. Instead, I think that we first need a clear account of what is going on at the banks on either side, as it were. The structural similarity that I have proposed offers preliminary hope that such bridges can be built.

The final topic that I wish to mention has to do with Husserl's own views on phenomenology and how it relates to empirical psychology. As discussed above, the intuiting of essences is central to Husserl's entire philosophical project, from the early *Logical Investigations* to his later transcendental idealism. Importantly, investigating essences is, for Husserl, entirely distinct from the inductive methods of the natural sciences.[10] Husserl took Hume's work very seriously; he regarded the inductive reasoning of natural science to be epistemically inferior to the strict descriptive methods of phenomenology, which is not inductive.

> The physical thing in which as a body a psyche is embodied, is as physical thing nothing at all but a unity of inductively belonging together ... it is merely a real unity arising from inductive causality. By no means does the parallelism of the psyche

which belongs to it signify that the psyche is also a merely real unity arising from inductive causality. (1968 § 23)

He goes on to suggest that regarding the psyche as a mere inductive unity, as an object in the natural world, stems from modern metaphysical prejudices tracing back to Descartes and Hobbes. This prejudice, very much alive today, holds that the method of the natural sciences is the "prototype" for all objective science (ibid.). Here we see some expression of Husserl's anti-naturalism, a theme developed more fully elsewhere in his work (1910/1965, 1976). I mention these views of Husserl's not to defend them here, but rather to introduce the speculative question of whether his own views on empirical science would be in conflict with the convergence that is the theme of this book.

I think that the key to appropriating Husserl's work for interdisciplinary purposes without coming into conflict with his own views lies in my limited explanatory scope and, relatedly, my explicit metaphysical neutrality. Husserl's misgiving about approaching the mind through natural science is based on the fact that natural science is essentially an inductive enterprise (while descriptive phenomenology is not). By indicating that there is a convergence between phenomenological structure, on one hand, and the structure of information processing in the embodied brain, on the other, I have not thereby violated Husserl's insistence on a noninductive approach to the phenomenal mind. We might use phenomenological insights as a way to make sense of what we find while engaged in inductive empirical research, but doing so should not conflict with Husserl's concerns above.[11]

I would like to conclude this appendix by noting that we are still in the early stages of conducting an interdisciplinary investigation into the mind that takes seriously the nature of first-person experience. The proper methodology remains an open question. My suggestion is that describing the abstract structure of different kinds of mental states, as pioneered by Husserl, is a promising method for engaging with results from the natural sciences, at least for the case of human visual experience.

Notes

Chapter 1

1. In the book I will follow the convention in the contemporary literature of using "content" to mean representational content, or, more precisely, an information state with correctness conditions (Peacocke 1983: 5). I address perceptual content in chapter 4.

2. Noë has since made it clear that he is not committed to the second strong claim (2010: 247).

3. One possible way to argue for a structure of anticipation and fulfillment in other modalities would be to give a kind of transcendental argument based on the possibility of surprise. Put roughly, anticipation might be a necessary condition for the possibility of being surprised. Surprise plays a role in my argument below, but I have not constructed a general argument based on surprise because I wish to emphasize the sensorimotor nature of visual anticipation, following Susanna Siegel's argument for perspectival connectedness (2010). See chapter 3 below.

4. Michael Tye (2000: chapter 4) also appeals to Marr's 2.5D sketch in an account of visual experience, but his view is that visual experience has multiple levels and is only partly determined by the 2.5D sketch. Also, Jean Petitot (1999) has pursued a strategy that is similar to mine, offering a synthesis of Husserl's description of vision with the cognitive neuroscience of vision. His technical treatment of the topic includes a brief discussion of perceptual anticipation (sections 7.2.2 and 7.2.3), but this is not his main theme. I will return to Petitot's work in sections 4.4 and A.3 in the appendix.

5. By "experience," I mean a conscious experience, something that refers to subjective awareness. In my terminology, "unconscious experience" is contradictory.

6. In Anderson's (2010) terminology, Prinz is advocating a version of anatomical modularity. Prinz (2006a) has been critical of the positing of modules, due to the Fodorian baggage of the term. Instead, he advocates functional decomposition into *subsystems* or *components*, not modules.

7. Egan and Matthews (2006) have cited these reasons as support for an approach to cognitive neuroscience known as dynamic causal modeling, developed by Karl Friston and colleagues (Friston, Harrison, and Penny 2003, Lee, Friston, and Horwitz 2006). Although I will not be discussing dynamic causal modeling as such, I will be appropriating related themes from Friston's work in chapters 5 and 6.

8. See http://www.humanconnectomeproject.org.

9. "We found that excitatory connectivity in cat area 17 is highly nonlocal; 74% of excitatory synapses near the axis of a 1,000-μm-diameter cortical column come from neurons located outside the column. Although we are not yet able to delineate the contributions from different sources to these long-range synapses, our results give a sense of how strongly interconnected cortical columns must be with one another, as well as with the structures of remote regions" (Stepanyants et al. 2009: 3558).

10. For a full discussion of this topic, with an emphasis on Dennett's influence on Hurley, see Madary (2012c, 2014a).

Chapter 2

1. The received view, which I accept, is that we do represent factual properties in visual experience. There have been recent challenges to this view; see section 4.3. For important challenges to the received view, see McDowell (1982), Travis (2013), and Brewer (2011). For a defense of the received view against these challenges, see Siegel (2010b) and Schellenberg (2011).

2. Using fMRI, Murray, Boyaci, and Kersten (2006) investigated whether neural activity in area V1 reflected the perspectival size or the factual size of visual stimuli. They reported that the neural activity corresponded to the factual size rather than the perspectival size. Since area V1 is a plausible location for a subpersonal representation of perspectival size, their results offer additional reasons to be skeptical of the existence such a representation.

3. Some may be reminded of the moon illusion with these experiments. The important difference here is that we can accurately represent the size of the stick visually, but the same is not true for the moon. When looking at celestial objects, perceptual constancy breaks down.

4. The way in which color experience depends on perspective has been treated in Husserl (1900/1993 V §2), Noë (2004: chapter 4), Thompson (2006), Cohen (2008), and Madary (2012a).

5. For some recent work on this issue, see Dainton (2000), Gallagher (2003), Phillips (2010), and Almäng (2014). It is an open question as to which current theory would fit best with AF, although a generally Husserlian approach is the most likely candidate.

6. Thanks to Ron Chrisley for discussing this point with me.

7. Fulfillments, as I will intend them here, are sensory. One could expand the notion of fulfillment systematically in order to include, say, aesthetic or ethical fulfillment. Such a move is obviously well-beyond my project here.

Chapter 3

1. See section 3.3 of Siegel (2010) for a discussion of the method of phenomenal contrast. In brief, it is a method of testing hypotheses about visual experience by considering (and contrasting) pairs of experiences. Siegel explains that the main strategy of this method "is to find something that the target hypothesis purports to explain, and then see whether it provides the best explanation of that phenomenon. ... Since contents of visual experiences are nonarbitrarily related to their phenomenal character, any target hypothesis will predict that certain pairs of such experiences that differ with respect to the hypothesized contents will differ phenomenally as well" (2010: 88).

2. Many who accept this view also accept that we represent a special kind of property that *does* depend on our particular location and viewing conditions. See section 2.1 above for a discussion of this view.

3. I do not suggest that all forms of surprise should be understood as violations of *perceptual* anticipations. In some cases, surprise is more naturally described as perceptual evidence that goes against one's belief. See my discussion of "stirred up" anticipations below for more details.

4. Although this argument is meant to stand without appeal to empirical results, I should mention well-known cases in which the "wrong" kinds of changes in visual phenomenology as a result of self-generated movement have occurred. In order to immerse subjects in virtual reality, the output of the head-mounted display must change in precisely the right ways in accordance with the self-generated movements of the subject. When the changes are off, such as when there is a lag between self-generated motion and the subsequent change in the visual display, subjects often experience a kind of motion sickness (Hettinger and Riccio 1992).

5. In defending weak representationalism, Michael Tye argues that there is nonconceptual representational content to an afterimage (2000: 85). It is plausible to read Tye's distinction between conceptual and nonconceptual content as mapping onto the distinction between factual and perspectival content, especially considering his discussion of the tilted coin (2000: 78–79). If this reading is correct, then it seems likely Tye would accept that there can be phenomenal experiences without *factual* content.

6. Thanks to Susanna Siegel for raising this issue.

7. Granting that there is self-identity over time for subjects.

8. This account of perceptual familiarity opens up a way to approach the question of whether we perceive higher-level properties, such as natural kinds, in perception (Siegel 2006, 2010; Prinz 2013; Logue 2013). Siegel has defended Thesis K, which states that "In some visual experiences, some K-properties are represented" (2006: 482). K-properties are meant to indicate properties beyond those that are typically thought to be represented in visual experience and include natural kind properties. I take the account I am developing to complement the concerns that motivate Siegel's claims in support of her Thesis K. In developing this line of thought further, I would suggest that visual anticipations can have content that is partly determined by higher-level properties such as natural kinds. See my discussion of visual content in chapters 4 and 9.

Chapter 4

1. For some of the core readings on perceptual content, see Jackson 1977, Searle 1983, Crane 1992, Peacocke 1992, Byrne 2001, Travis 2004, Martin 2006, Siegel 2010b, and Schellenberg 2011.

2. To be clear, I am suggesting that the philosophy of perception should not be based on the philosophy of language. I am not suggesting that linguistic reports cannot illuminate features of visual experience (see Mulligan 1999).

3. For a detailed investigation into the distinction between perceptual and propositional content from a Husserlian perspective, see Smith and McIntyre (1982).

4. I am not implying that Schwitzgebel is committed to the view that perceptual content is propositional.

5. I return to this theme in the appendix when discussing Tim Bayne's work on introspection (2004, 2015).

6. Beginning with Rorty in expressing this worry is not meant to imply that its lineage begins there. The origins go back at least to Sellars (1956/1997), whose work on the topic was influenced by his reading of Hegel.

7. In his more recent work, McDowell, influenced by Charles Travis (2004), has rejected the view that experiences have propositional content, although he still thinks that experiential content is conceptual (2009: 258–264). I will not cover how the views of McDowell and Travis might fit with the view I am proposing, nor will I comment on whether McDowell's new view can overcome the Davidson/McDowell worry. For a discussion of Husserl and McDowell on related issues, see Doyon (2011).

Notes

Chapter 5

1. For other places in which sensorimotor theorists express their doubts about the explanatory usefulness of visual neuroscience, see O'Regan and Noë 2001, Noë 2004, O'Regan 2011, and O'Regan and Block 2012. Also see my review of Noë's *Varieties of Presence* (Madary 2015b) for further discussion of this issue.

2. I consider my view to be in broad agreement with Seth's (2014) position, though see Madary (2014b) for some details of where our positions diverge.

3. I return to this theme in section 8.1.

4. I am not suggesting that Pylyshyn is unaware of peripheral indeterminacy, or of some of the other reasons given for rejecting the myth of full detail. In discussing such matters, Pylyshyn seeks to answer the question of how we manage to see a panorama of full detail despite the limitations of impoverished retinal processing (2003: 7–8). I reject his question because I reject the assumption that we see a panorama of full detail.

5. In her discussion of the topic, Susan Blackmore (2002: 22) has framed the debate in terms of whether or not change blindness is compatible with positing internal representations. I will discuss this topic below in section 8.3.

6. Another kind of evidence commonly cited in support of the connection between action and vision is sensory substitution (Bach-y-Rita 1996, O'Regan and Noë 2001). I make no appeal to sensory substitution because of difficult issues about how to interpret the results (Deroy and Auvray, 2014).

7. For much more detail, see Findlay and Gilchrist (2003, especially chapter 2).

8. Early studies on left-right distorting goggles were carried out by James Taylor (1962). For a recent version of the experiment, see Jan Degenaar (2013).

9. For recent work in this area, see Aline Bompas and Kevin O'Regan (2006).

10. In my view, "visual feedback" is a subpersonal description, and "visual anticipation" is a personal-level description. For now, suffice it to say that the two are closely related. My understanding of the relationship between the two will be addressed in chapter 8.

11. Though I should note that the predictive-processing framework, which fits nicely with my general claims, may be able to avoid the binding problem that the feature-integration theory is trying to solve. Hohwy writes that "the system does not have to operate in a bottom-up fashion and first process individual attributes and then bind them. Instead, it assumes bound attributes and then predicts them down through the cortical hierarchy" (2012: 6).

12. A high indeterminacy of unconsciously processed stimuli is compatible with empirical results that indicate sophisticated "high-level" processing of those stimuli. Even, for example, semantic priming need not involve a determinate representation in order to have a measurable effect.

Chapter 6

1. In what follows I assume that the reader has some basic knowledge of visual neuroscience. For introductory texts, see Palmer (1999) or Goldstein (2002, or other editions).

2. For an entire book on the theme, see Buzsaki (2006).

3. See Pinel (2003, chapter 4) for a standard textbook account of neural transmission.

4. For a seminal paper in this area, see Rinzel and Ermentrout (1989). For an example of more recent work, see Nawrot et al. (2008). For a recent comprehensive introduction, see Izhikevich (2007).

5. The local field potential measures electrical activity in an extracellular region of cortex and not from any particular neuron in that region.

6. For subsequent work on this topic, see Kenet et al. 2003, Leopold, Murayama, and Logothetis 2003, Petersen et al. 2003, Parga and Abbott 2007, and Han, Caporale, and Dan 2008.

7. See the articles in the special issue of *Frontiers in Psychology* (Sharp and Leech 2012) for a sample of contemporary debate.

8. As in Prinz: "Indeed, there is evidence that such feedback is anatomically limited and functionally variable; this suggests that feedback plays only a modest and context-sensitive role in modulating inputs to reduce conflict or reflect visual expectation" (2012: 72).

9. For more details about the way in which predictive processing might be implemented at the neural level, see Bastos et al. (2012).

10. Some of the early articles introducing generative models include Zemel (1993), Zemel and Hinton (1995) and Hinton et al. (1995). For a textbook introduction, see Dayan and Abbott (2005, chapter 10). For more recent, and more accessible, overviews of the concept, see Hinton (2006, 2007).

11. For clear expositions of this approach from a philosophical perspective, see Andy Clark (2013b) and Jakob Hohwy (2013). One of the figures central to developing this approach has been Karl Friston (2005, 2008, 2010).

Notes

12. For some attempts to explain extra-classical field effects without predictive processing, see Dobbins, Zucker, and Cynader (1987), Peterhans and von der Heydt (1991), and Grossberg, Mingolla, and Ross (1997).

Chapter 7

1. For some examples of Milner and Goodale's main hypothesis influencing discussions of visual consciousness, see Crick and Koch (1998: 98) and Chalmers (2000: 21). Also note that some of Andy Clark's work on this topic supposes that Milner and Goodale are correct in emphasizing the dichotomy between conscious vision for perception versus unconscious vision for action. In his influential article from 2001, for example, both the assumption of experience-based control and the hypothesis of experience-based selection are formulated without mention of the temporal and spatial scales at play in conscious experience.

2. I should mention one important exception. In a recent article, Matthieu De Wit and colleagues (de Wit, van der Kamp, and Masters 2012) have proposed an interpretation of the empirical evidence that is quite similar to what I am suggesting here.

3. For some recent philosophical work that challenges the dichotomy, see Noë (2004), Schellenberg (2008), and Briscoe (2009). Of course, the Gibsonian ecological tradition in psychology could also motivate criticism of the dichotomy (see section 5.1.).

4. I do not mean to suggest that the visual horizon is limited to the visual periphery. It is in the nature of horizons, I take it, that they cannot be determinately located in space. Thanks to Kristjan Laasik (2014) for critically discussing this point.

5. My comments here are about the physiological nature of the inputs to the two streams, and I am not going to enter the somewhat large debate over egocentric versus allocentric *coding* in the two streams. For a treatment of this issue from a philosophical perspective, see Briscoe (2009).

6. There is also a third koniocellular pathway that is not as well understood as the other two major pathways. The koniocellular pathway includes far fewer cells than the other two (Kveraga 2007, Callaway 2005).

7. For a review of the challenges to their proposal, see Milner and Goodale (1995: 34–36). Some important articles on this topic include Schiller and Logothetis (1990), Merigan and Maunsell (1993), and, more recently, Nassi and Callaway (2006, 2009).

8. For an alternative view based on macaque retina, see Silveira and Perry (1991).

9. Laasik has noted that D.F.'s ability to grasp after *some* delay, even a delay of less than two seconds, suggests that it is not anticipations that are relevant. He claims that the relevant explanation "has to do with the retention of information in

memory, perhaps similar to the Husserlian retention" (2014: 434). I thank him for catching this point. My reply is that stirred up visual anticipations are transient, and that the anticipations enabled by the dorsal stream fade away faster than those enabled by the ventral stream. D.F.'s ability can be explained by anticipations being retained just long enough to complete the action after a very short delay.

10. Jacob and Jeannerod also discuss patient S.B. (2003: 88–89).

11. In what follows, I partly base my usage of these terms on Dubois and VanRullen (2011), who cite Bender (1945), Meadows and Munro (1977), Zeki (1991), Norton and Corbett (2000), and Horton and Trobe (2009), among others, as the basis for their definitions.

12. There is at least one instance of a disturbance of self-generated motion for visual perception in the literature, but this case does not involve reports of snapshot visual experiences, as in akinetopsia. Patient R.W. complained of vertigo and nausea when tracking externally moving objects through self-generated eye movements (watching children on a playground) and when tracking stationary objects while moving relative to them (from inside a moving car, for instance). Haarmeier et al. (1997) determined that R.W. suffered from a selective impairment of the ability to perceive a stationary world during self-generated eye movements. The cause of this impairment is thought to be bilateral extrastriate lesions (near, and perhaps including a part of, MT/V5) incurred in early childhood.

13. I have not found reports of this kind of experiment in the literature on akinetopsia, which is not surprising given the scarcity of subjects with the disorder.

14. Though I should be clear that the precise location of L.M.'s damage relative to MT/V5 is not known, see Hess et al. (1989) for a discussion.

15. As the previous chapters should make clear, I am not suggesting here that visual anticipation is exclusively enabled by the dorsal stream. Thanks to Nivedita Gangopadhyay for questioning me on this point.

Chapter 8

1. See Costall (2006) for a treatment of this topic that seeks to debunk some of the purported misconceptions about the history of psychology.

2. Dennett has claimed that there is some misunderstanding of his view (2003). For various perspectives on Dennett's work on this topic, see the special issue of *Phenomenology and the Cognitive Sciences* edited by Gallagher and Zahavi (2007).

3. Here I have in mind thinkers such as Maurice Merleau-Ponty, Francisco Varela, Susan Hurley, Shaun Gallagher, Thomas Metzinger, Alva Noë, Dan Zahavi, and Evan Thompson, just to name a few. In many ways, my strategy follows a strategy that has become known as "naturalizing phenomenology" (Petitot et al. 1999) or "neuro-

phenomenology" (Varela 1996), though see the appendix for important ways in which my project diverges from this strategy.

4. For an explicitly Husserlian application of structural coherence between phenomenology and neural models, see Yoshimi (2011).

5. In section 3.4, I suggest that visual anticipations may be a nonstandard instance of personal-level content.

6. Here I allow for the possibility of describing perceptual states using propositional attitudes and natural language, but I remain committed to the view developed in chapter 4 that this is not the best way to understand perceptual content.

7. According to some views on the topic, the content of mental representations has to do with whether stimuli are relevant for the biological needs of the organism; these views are called "deflationary" by Burge (2010: 293). The view that I am urging goes beyond the biological relevance of the objective stimulus to include contextually sensitive features of the organism and its ongoing interaction with its environment, as in Thompson (2007: chapter 3).

8. There is also the possibility of a system showing anticipatory behavior without having an internal model. Dubois (2003) has referred to anticipation with and without an internal model as "weak" and "strong" anticipation, respectively. Stephen et al. (2008) and Stephen and Dixon (2011) have suggested that there may be strongly anticipatory cognitive systems, in which there is anticipation without an internal model. If they are correct, and here I must leave it an open (and fascinating) question, then we could have personal-level anticipatory visual content without a subpersonal internal model.

9. As I have noted elsewhere (Madary 2015a), this idea of organism-relative content finds expression in a number of different places, including von Uexküll's *Umwelt* (1934), Merleau-Ponty's (1962) discussion of sensory stimuli (which is a central influence on contemporary enactivism [Thompson 2007]), Millikan's "pushmi-pullyu" representations (1995), Akins' narcissistic sensory systems (1996), Clark's earlier work on embodied cognition (1997: chapter 1), and Metzinger's ego tunnel (2009: 8–9).

Chapter 9

1. See Anthony Steinbock (1995, especially pages 132–137) for a discussion of Husserl's treatment of closely related themes.

2. Also highly relevant here is Colwyn Trevarthen's work on primary intersubjectivity (1979).

3. An excellent demonstration can be found at www.biomotionlab.ca.

4. For further work on the anticipatory nature of social vision, see Adams and Kveraga (2015). For a review of the literature on predictive processing and social cognition, see Brown and Brüne (2012). Finally, approaches to social interaction in terms of synchronization (Konvalinka et al. 2011, Friston and Frith 2015) also show support for VCS due to the fact that the content of visual anticipations is sensitive to temporal properties of the perceptual scene.

5. To be even more parsimonious, one could even look for a single computational strategy underlying both perception and action, thus putting pressure on the distinction itself. I will not defend this strategy, but for a step in this direction, see Clark (2013b, 2014), who is exploring themes from Karl Friston's work.

Appendix

1. For an excellent overview of Husserl's theory of perception that goes beyond my focus here, see Mulligan (1995).

2. Compare Noë's remarkably similar claim: "When you look at the wall, you see its uniform color *in* its evident variation in color across its surface" (2004: 166).

3. For a closer look at Husserl's earlier struggles with deciding on the proper terminology, see Husserl (2004: 144–145). I am following Rodemeyer in identifying perceptual anticipations as a kind of protention: "Temporalizing consciousness of a 'future' is not merely an empty anticipation of the distant future, nor is it an empty forward-movement into what might become a fulfilled impression. Instead, *what is expected is part of what is intended and fulfilled*—it is part of my experience" (2006: 135).

4. One interesting, yet relatively unexplored, philosophical issue has to do with the degree to which the seeds of transcendental idealism can be found in the *Logical Investigations*, prior to Husserl's turn. As a matter of historical fact, Husserl's students at the time, as well as Husserl himself, regarded the turn as a major break in his philosophical trajectory. Herbert Spiegelberg reports that his Göttingen students "responded with growing consternation" in 1907 (1978 vol. 1: 170). Still, there are elements of continuity, such as Husserl's emphasis on essences and, importantly for our purposes, his description of the structure of perceptual experience. There are also bits of the *Logical Investigations* that may appear, to the contemporary reader, to advocate anti-realism about the mind-independent world, such as "Everything that is, can be known 'in itself'" (1900/1993 I §28). This claim is quite similar to the position that Timothy Williamson (1982) refers to as "weak anti-realism," a position that has received a great deal of consideration in the subsequent literature on Fitch's Paradox.

5. For the case that it is a rather strong form of idealism, see A. D. Smith (2003). For a fascinating discussion of Husserl's transcendental idealism as it relates to issues in philosophy of mind and analytic philosophy more generally, see Zahavi (2008).

6. For a reading of Husserl's notion of constitution that does not conflict with the ambivalence about transcendental idealism in the present text, see Yoshimi (2016).

7. It is interesting to note that Burge, motivated entirely by perceptual psychology, may be in agreement with Husserl here. He "conjecture[s]" that perceptual constancy is necessary for "perception and objectivity" (2010: 413). To be clear, though, Burge has his own misgivings about Husserl's methodology (Burge 2010: 130–133).

8. For a reference work on this theme, see Schmicking and Gallagher (2010).

9. An accessible discussion of Husserl's method of intuiting essences can be found in his lectures on *Phenomenological Psychology* (1968, especially §9).

10. Though Husserl's own position has been challenged on this point (Levin 1968).

11. On this topic, Evan Thompson, following both Varela and Merleau-Ponty, has claimed that contemporary models of neural dynamics, models with irreducibly relational structures and internal teleology, do in fact assuage some of Husserl's anti-naturalistic worries (2007: 357). Since I have not adopted these models in particular, I will not discuss this proposal in detail. From what I can tell, though, Husserl's more general worry about the difference between investigating essences, on one hand, and investigating the natural world through inductive methods of the natural sciences, on the other, is one that is not overcome with the kinds of models that Thompson and other enactivists have adopted.

References

Abeles, M. (1982). *Local cortical circuits: An electrophysiological study*. Studies of brain function (Vol. 6). Berlin: Springer-Verlag.

Abeles, M., & Prut, Y. (1996). Spatio-temporal firing patterns in the frontal cortex of behaving monkeys. *Journal of Physiology (Paris), 90*(3–4), 249–250.

Abert, B., & Ilsen, P. F. (2010). Palinopsia. *Optometry, 81*(8), 394–404. doi:10.1016/j.optm.2009.12.010.

Adams, F., & Aizawa, K. (2008). *The bounds of cognition*. Malden, MA: Blackwell.

Adams, R. B., & Kveraga, K. (2015). Social vision: Functional forecasting and the integration of compound social cues. *Review of Philosophy and Psychology, 6*(4), 591–610. doi:10.1007/s13164-015-0256-1.

Aglioti, S., DeSouza, J. F., & Goodale, M. A. (1995). Size-contrast illusions deceive the eye but not the hand. *Current Biology, 5*(6), 679–685. doi:10.1016/S0960-9822(95)00133-3.

Aizawa, K. (2007). Understanding the embodiment of perception. *Journal of Philosophy, 104*(1), 5–25. doi:10.2307/20619993.

Akins, K. (1996). Of sensory systems and the "aboutness" of mental states. *Journal of Philosophy, 93*(7), 337. doi:10.2307/2941125.

Allman, J., Miezin, F., & McGuinness, E. (1985). Stimulus specific responses from beyond the classical receptive field: Neurophysiological mechanisms for local-global comparisons in visual neurons. *Annual Review of Neuroscience, 8*, 407–430. doi:10.1146/annurev.ne.08.030185.002203.

Almäng, J. (2014). Tense as a feature of perceptual content. *Journal of Philosophy, 111*(7), 361–378.

Alving, B. O. (1968). Spontaneous activity in isolated somata of aplysia pacemaker neurons. *Journal of General Physiology, 51*(1), 29–45. doi:10.1085/jgp.51.1.29.

Anderson, M. L. (2007). The massive redeployment hypothesis and the functional topography of the brain. *Philosophical Psychology, 21*, 143–174.

Anderson, M. L. (2010). Neural reuse: A fundamental organizational principle of the brain. *Behavioral and Brain Sciences, 33*(04), 245–266. doi:10.1017/S0140525X10000853.

Anderson, M. L. (2014). *After phrenology: Neural reuse and the interactive brain.* Cambridge: MIT Press.

Anderson, M., Brumbaugh, J., & Suben, A. (2010). Investigating functional cooperation in the human brain using simple graph-theoretic methods. In Chaovalitwongse, Pardalos, & Xanthopoulos, *Computational neuroscience*, 31–41.

Angelaki, D. E., & Cullen, K. E. (2008). Vestibular system: The many facets of a multimodal sense. *Annual Review of Neuroscience, 31*, 125–150. doi:10.1146/annurev.neuro.31.060407.125555.

Arend, L., & Reeves, A. (1986). Simultaneous color constancy. *Journal of the Optical Society of America. A, Optics and Image Science, 3*(10), 1743–1751.

Arend, L. E., Reeves, A., Schirillo, J., & Goldstein, R. (1991). Simultaneous color constancy: Paper with diverse Munsell values. *Journal of the Optical Society of America. A, Optics and Image Science, 8*(4), 661–672.

Arieli, A., Sterkin, A., Grinvald, A., & Aertsen, A. (1996). Dynamics of ongoing activity: Explanation of the large variability in evoked cortical responses. *Science, 273*(5283), 1868–1871.

Armstrong, D. M. (1968). *A materialist theory of the mind.* International library of philosophy and scientific method. London: Routledge.

Armstrong, D. M. (1999). *The mind-body problem: An opinionated introduction.* Focus series. Boulder, Colo.: Westview Press.

Armstrong, K. M., & Moore, T. (2007). Rapid enhancement of visual cortical response discriminability by microstimulation of the frontal eye field. *Proceedings of the National Academy of Sciences of the United States of America, 104*(22), 9499–9504. doi:10.1073/pnas.0701104104.

Arvanitaki, A. (1939). Recherches sur la réponse oscillatoire locale de l'axone géant isolé de « sepia ». *Archives of Physiology and Biochemistry, 49*(2), 209–256. doi:10.3109/13813453909150823.

Atmanspacher, H. and beim Graben, P. (2007). Contextual emergence of mental states from neurodynamics. *Chaos and Complexity Letters, 2*(2–3), 151–168.

Ayer, A. J. (1952). *Language, truth, and logic.* New York: Dover.

References

Bach-y-Rita, P. (2002). Sensory substitution and qualia. In Noë & Thompson, *Vision and mind*, 497–514.

Backus, B. T., Fleet, D. J., Parker, A. J., & Heeger, D. J. (2001). Human cortical activity correlates with stereoscopic depth perception. *Journal of Neurophysiology*, 86(4), 2054–2068.

Bain, A. (2006). *The emotions and the will*. Cosimo classics. New York: Cosimo. (1888).

Ballard, D. H., Hayhoe, M. M., & Pelz, J. B. (1995). Memory representations in natural tasks. *Journal of Cognitive Neuroscience*, 7(1), 66–80. doi:10.1162/jocn.1995.7.1.66.

Ballard, D. H., Hayhoe, M. M., Pook, P. K., & Rao, R. P. (1997). Deictic codes for the embodiment of cognition. *Behavioral and Brain Sciences*, 20(4), 723–742, discussion 743–767.

Bar, M. (2003). A cortical mechanism for triggering top-down facilitation in visual object recognition. *Journal of Cognitive Neuroscience*, 15(4), 600–609. doi:10.1162/089892903321662976.

Bar, M. (2007). The proactive brain: Using analogies and associations to generate predictions. *Trends in Cognitive Sciences*, 11(7), 280–289. doi:10.1016/j.tics.2007.05.005.

Barber, M. (2008). Holism and horizon: Husserl and McDowell on non-conceptual content. *Husserl Studies*, 24(2), 79–97.

Barrett, L. F., & Bar, M. (2009). See it with feeling: Affective predictions during object perception. *Philosophical Transactions of the Royal Society of London. Series B, Biological Sciences*, 364(1521), 1325–1334. doi:10.1098/rstb.2008.0312.

Bassett, D. S., & Bullmore, E. (2006). Small-world brain networks. *The Neuroscientist* 12 (6), 512–523. doi:10.1177/1073858406293182.

Bastos, A. M., Usrey, W. M., Adams, R. A., Mangun, G. R., Fries, P., & Friston, K. J. (2012). Canonical microcircuits for predictive coding. *Neuron*, 76(4), 695–711. doi:10.1016/j.neuron.2012.10.038.

Bayne, T. (2015). Introspective insecurity. In Metzinger & Windt, *Open mind: Philosophy and the mind sciences in the 21st century*.

Bayne, T. (2004). Closing the gap? Some questions for neurophenomenology. *Phenomenology and the Cognitive Sciences*, 3(4), 349–364. doi:10.1023/B:PHEN.0000048934.34397.ca.

Bechtel, W. (1998). Representations and cognitive explanations: Assessing the dynamicist's challenge in cognitive science. *Cognitive Science*, 22(3), 295–318. doi:10.1207/s15516709cog2203_2.

Bechtel, W. (2013). The endogenously active brain: The need for an alternative architecture. *Philosophy of Science, 17*, 3–30.

Beck, J. (Ed.). (1982). *Organization and representation in perception*. Hillsdale, NJ: Lawrence Erlbaum.

Behrmann, M., Thomas, C., & Humphreys, K. (2006). Seeing it differently: Visual processing in autism. *Trends in Cognitive Sciences, 10*(6), 258–264. doi:10.1016/j.tics.2006.05.001.

Belot, M., Crawford, V. P., & Heyes, C. (2013). Players of Matching Pennies automatically imitate opponents' gestures against strong incentives. *Proceedings of the National Academy of Sciences of the United States of America, 110*(8), 2763–2768. doi:10.1073/pnas.1209981110.

Bender, M. B. (1945). Polyopia and monocular diplopia of cerebral origin. *Archives of Neurology and Psychiatry, 54*, 323–338.

Benson, D., & Greenberg, J. (1969). Visual form agnosia. *Archives of Neurology, 20*, 82–89.

Bernet, R., Kern, I., & Marbach, E. (1993). *An introduction to Husserlian phenomenology*. Evanston, IL: Northwestern University Press.

Bickhard, M. (2002). Mind as a process. In Riffert & Weber, *Searching for new contrasts*, 285–294.

Binzegger, T., Douglas, R. J., & K. A. C. Martin (2004). A quantitative map of the circuit of cat primary visual cortex. *Journal of Neuroscience, 24*(39), 8441–8453. doi:10.1523/JNEUROSCI.1400-04.2004.

Blackmore, S. (2002). There is no stream of consciousness. *Journal of Consciousness Studies, 9*(5), 17–28.

Blake, A., & Bülthoff, H. (1990). Does the brain know the physics of specular reflection? *Nature, 343*(6254), 165–168. doi:10.1038/343165a0.

Block, N. (2003). Mental paint. In Hahn & Ramberg, *Reflections and replies*, 165–200.

Block, N. (Ed.). (1981). *Readings in philosophy of psychology*. Cambridge, Mass.: Harvard University Press.

Block, N. (2005). Review of Alva Noë, Action in Perception. *Journal of Philosophy, CII*(5), 259–272.

Block, N. (2007). Consciousness, accessibility, and the mesh between psychology and neuroscience. *Behavioral and Brain Sciences, 30*(5), 481–548.

Block, N. (2011). Perceptual consciousness overflows cognitive access. *Trends in Cognitive Sciences, 15*(12), 567–575. doi:10.1016/j.tics.2011.11.001.

References

Block, N. (2014). Rich conscious perception outside focal attention. *Trends in Cognitive Sciences, 18*(9), 445–447. doi:10.1016/j.tics.2014.05.007.

Bock, D. D., Lee, W.-C. A., Kerlin, A. M., Andermann, M. L., Hood, G., Wetzel, A. W., et al. (2011). Network anatomy and in vivo physiology of visual cortical neurons. *Nature, 471*(7337), 177–182. doi:10.1038/nature09802.

Boduroglu, A., Shah, P., & Nisbett, R. E. (2009). Cultural differences in allocation of attention in visual information processing. *Journal of Cross-Cultural Psychology, 40*(3), 349–360. doi:10.1177/0022022108331005.

Bompas, A., & O'Regan, J. K. (2006). Evidence for a role of action in colour perception. *Perception, 35*(1), 65–78. doi:10.1068/p5356.

Braitenberg, V., & Schüz, A. (1998). *Cortex: Statistics and geometry of neuronal connectivity* (Rev. ed.). Berlin: Springer.

Brewer, B. (2006). Perception and content. *European Journal of Philosophy, 14*(2), 165–181.

Brewer, B. (2011). *Perception and its objects*. Oxford: Oxford University Press.

Breyer, T. (2011). *Attentionalität und Intentionalität: Grundzüge einer phänomenologisch-kognitionswissenschaftlichen Theorie der Aufmerksamkeit. Phänomenologische Untersuchungen 28*. München: Fink.

Briscoe, R. (2008). Vision, action, and make-perceive. *Mind & Language, 23*(4), 457–497.

Briscoe, R. (2009). Egocentric spatial representation in action and perception. *Philosophy and Phenomenological Research, 79*(2), 423–460. doi:10.1111/j.1933-1592.2009.00284.x.

Brockman, J. 1996. *The third culture*. New York: Simon & Schuster.

Brogaard, B. (2010). Strong representationalism and centered content. *Philosophical Studies: An International Journal for Philosophy in the Analytic Tradition, 151*(3), 373–392.

Bronfman, Z. Z., Brezis, N., Jacobson, H., & Usher, M. (2014). We see more than we can report: "Cost free" color phenomenality outside focal attention. *Psychological Science, 25*(7), 1394–1403. doi:10.1177/0956797614532656.

Brown, E. C., & Brüne, M. (2012). The role of prediction in social neuroscience. *Frontiers in Human Neuroscience, 6*, 147. doi:10.3389/fnhum.2012.00147.

Brown, L. E., Halpert, B. A., & Goodale, M. A. (2005). Peripheral vision for perception and action. *Experimental Brain Research, 165*(1), 97–106. doi:10.1007/s00221-005-2285-y.

Bruner, J., & Goodman, C. (1947). Value and need as organizing factors in perception. *Journal of Abnormal and Social Psychology, 42*, 33–44.

Bruner, J., & Postman, L. (1949). On the perception of incongruity: A paradigm. *Journal of Personality, 18*, 206–223.

Bullier, J. (2001a). Feedback connections and conscious vision. *Trends in Cognitive Sciences, 5*(9), 369–370. doi:10.1016/S1364-6613(00)01730-7.

Bullier, J. (2001b). Integrated model of visual processing. *Brain Research. Brain Research Reviews, 36*(2–3), 96–107. doi:10.1016/S0165-0173(01)00085-6.

Bullowa, M. (Ed.). (1979). *Before speech: The beginning of interpersonal communication*. Cambridge, UK: Cambridge University Press.

Burge, T. (2010). *Origins of objectivity*. Oxford: Oxford University Press.

Butz, M. V., Sigaud, O. & Gérard, P. (Eds.). (2003). *Anticipatory behavior in adaptive learning systems: Foundations, theories, and systems*. Lecture notes in computer science 2684. Berlin: Springer.

Buzsáki, G. (2006). *Rhythms of the brain*. Oxford: Oxford University Press.

Byrne, A. (2001). Intentionalism defended. *Philosophical Review, 110*(2), 199–240.

Callaway, E. M. (2005). Structure and function of parallel pathways in the primate early visual system. *Journal of Physiology, 566*(Pt 1), 13–19. doi:10.1113/jphysiol.2005.088047.

Carman, T., & Hansen, M. (Eds.). (2004). *The Cambridge Companion to Merleau-Ponty*. Cambridge University Press.

Carruthers, P. (2005). *Consciousness: Essays from a higher-order perspective*. Oxford: Oxford University Press.

Carruthers, P. (2009). How we know our own minds: The relationship between mindreading and metacognition. *Behavioral and Brain Sciences, 32*(2), 121–138, discussion 138–182. doi:10.1017/S0140525X09000545.

Carsetti, A. (Ed.). (2010). *Functional models of cognition: Self-organizing dynamics and semantic structures in cognitive systems*. Theory and decision library. Series A 27. Dordrecht: Kluwer.

Casler, K., Terziyan, T., & Greene, K. (2009). Toddlers view artifact function normatively. *Cognitive Development, 24*(3), 240–247. doi:10.1016/j.cogdev.2009.03.005.

Chalmers, D. (1995). Facing up to the problem of consciousness. *Journal of Consciousness Studies, 2*(3), 200–219.

Chalmers, D. (2000). What is a neural correlate of consciousness? In Metzinger, *Neural correlates of consciousness*, 17–40.

References

Chalmers, D. J. (1996). *The conscious mind: In search of a theory of conscious experience.* Philosophy of mind series. New York: Oxford University Press.

Chalupa, L. M., & Werner, J. S. (Eds.). (2004). *The visual neurosciences.* Cambridge, Mass.: MIT Press.

Chaovalitwongse, W., Pardalos, P., & Xanthopoulos, P. (Eds.). (2010). *Computational neuroscience.* New York: Springer.

Chemero, A. (2009). *Radical embodied cognitive science.* Cambridge, Mass.: MIT Press.

Chisholm, R. (1942). The problem of the speckled hen. *Mind, 51*(204), 368–373.

Churchland, P. (1981). Eliminative materialism and the propositional attitudes. *Journal of Philosophy, LXXVIII*(2), 67–90.

Churchland, P. S., Ramachandran, V. S., & Sejnowski, T. (1994). A critique of pure vision. In Koch & Davis, *Large-scale neuronal theories of the brain,* 23–60.

Clark, A. (1997). *Being there: Putting brain, body, and world together again.* Cambridge, Mass.: MIT Press.

Clark, A. (2001). Visual experience and motor action: Are the bonds too tight? *Philosophical Review, 110*(4), 495–519. doi:10.1215/00318108-110-4-495.

Clark, A. (2013a). The many faces of precision (Replies to commentaries on Whatever next? Neural prediction, situated agents, and the future of cognitive science). *Frontiers in Psychology, 4,* 270. doi:10.3389/fpsyg.2013.00270.

Clark, A. (2013b). Whatever next? Predictive brains, situated agents, and the future of cognitive science. *Behavioral and Brain Sciences, 36*(3), 181–204. doi:10.1017/S0140525X12000477.

Clark, A. (2015a). *Surfing uncertainty: Prediction, action, and the embodied mind.* New York: Oxford University Press.

Clark, A. (2015b). Predicting peace. In Metzinger & Windt, *Open mind: Philosophy and the mind sciences in the 21st century.*

Cohen, J. (2008). Colour constancy as counterfactual. *Australasian Journal of Philosophy, 86*(1), 61–92.

Cohen, M. A., & Dennett, D. C. (2011). Consciousness cannot be separated from function. *Trends in Cognitive Sciences, 15*(8), 358–364. doi:10.1016/j.tics.2011.06.008.

Cohen, G., Johnston, R. A., & Plunkett, K. (Eds.). (2000). *Exploring cognition: Damaged brains and neural networks.* Readings in Cognitive Neuropsychology and Connectionist Modelling. Sussex: Psychology Press.

Cohen, M. R., & Newsome, W. T. (2008). Context-dependent changes in functional circuitry in visual area MT. *Neuron, 60*(1), 162–173. doi:10.1016/j.neuron.2008.08.007.

Colby, C. L., Gattass, R., Olson, C. R., & Gross, C. G. (1988). Topographical organization of cortical afferents to extrastriate visual area PO in the macaque: A dual tracer study. *Journal of Comparative Neurology, 269*(3), 392–413. doi:10.1002/cne.902690307.

Cornelissen, F. W., & Brenner, E. (1995). Simultaneous colour constancy revisited: An analysis of viewing strategies. *Vision Research, 35*(17), 2431–2448.

Costall, A. (2006). "Introspectionism" and the mythical origins of scientific psychology. *Consciousness and Cognition, 15*(4), 634–654. doi:10.1016/j.concog.2006.09.008.

Covic, E. N., & Sherman, S. M. (2011). Synaptic properties of connections between the primary and secondary auditory cortices in mice. *Cerebral Cortex, 21*(11), 2425–2441. doi:10.1093/cercor/bhr029.

Crane, T. (Ed.). (1992). *The contents of experience*. Cambridge, UK: Cambridge University Press.

Crane, T. (Ed.). (2009). Is perception a propositional attitude? *Philosophical Quarterly, 59*(236), 452–469.

Crick, F., & Koch, C. (1998). Constraints on cortical and thalamic projections: The no-strong-loops hypothesis. *Nature, 391*(6664), 245–250. doi:10.1038/34584.

Cummins, R. (1989). *Meaning and mental representation*. Cambridge, Mass.: MIT Press.

Cutsuridis, V., Hussain, A., & Taylor, J. G. (2011). *Perception-action cycle: Models, architectures, and hardware*. Biomedical and life sciences (Vol. 1). New York: Springer.

Dacey, D. M., & Petersen, M. R. (1992). Dendritic field size and morphology of midget and parasol ganglion cells of the human retina. *Proceedings of the National Academy of Sciences of the United States of America, 89*(20), 9666–9670. doi:10.1073/pnas.89.20.9666.

Dainton, B. (2000). *Stream of consciousness: Unity and continuity in conscious experience*. International library of philosophy. London: Routledge.

Dale, R., & Spivey, M. J. (2005). From apples and oranges to symbolic dynamics: A framework for conciliating notions of cognitive representation. *Journal of Experimental & Theoretical Artificial Intelligence, 17*(4), 317–342. doi:10.1080/09528130500283766.

Davidson, D. (2001). *Subjective, intersubjective, objective*. Oxford: Oxford University Press.

Davies, M. (2010). Double dissociation: Understanding its role in cognitive neurophysiology. *Mind & Language, 25*(5), 500–540.

References

Dayan, P. (1995). The Helmholtz machine. *Neural Computation, 7*, 889–904.

Dayan, P., & Abbott, L. F. (2001). *Theoretical neuroscience: Computational and mathematical modeling of neural systems. Computational neuroscience.* Cambridge, Mass.: MIT Press.

De Wit, M. M., van der Kamp, J., & Masters, R. S. W. (2012). Distinct task-independent visual thresholds for egocentric and allocentric information pick up. *Consciousness and Cognition, 21*(3), 1410–1418. doi:10.1016/j.concog.2012.07.008.

Degenaar, J. (2014). Through the inverting glass: First-person observations on spatial vision and imagery. *Phenomenology and the Cognitive Sciences, 13*(2), 373–393. doi:10.1007/s11097-013-9305-3.

Dennett, D. C. (1969). *Content and consciousness.* London: Routledge.

Dennett, D. C. (1987). *The intentional stance.* Cambridge, Mass: MIT Press.

Dennett, D. C. (1991). *Consciousness explained* (1st ed.). Boston: Little, Brown.

Dennett, D. C. (2001). Surprise, surprise. *Behavioral and Brain Sciences, 24*, 982.

Dennett, D. C. (2003). Who's on first? Heterophenomenology explained. *Journal of Consciousness Studies, 10*(9), 19–30.

Depraz, N. (2004). Where is the phenomenology of attention that Husserl intended to perform? A transcendental pragmatic-oriented description of attention. *Continental Philosophy Review, 37*(1), 5–20. doi:10.1023/B:MAWO.0000049309.87813.f7.

Deroy, O., & Auvray, M. (2014). Beyond vision: The vertical integration of sensory substitution devices. In Stokes, Matthen, & Biggs, *Perception and its modalities*, 473–476).

Ditchburn, R. W., & Ginsborg, B. L. (1952). Vision with a stabilized retinal image. *Nature, CLXX*, 36–37.

Dittrich, W. H., Troscianko, T., Lea, S. E., & Morgan, D. (1996). Perception of emotion from dynamic point-light displays represented in dance. *Perception, 25*(6), 727–738.

Dobbins, A., Zucker, S. W., & Cynader, M. S. (1987). Endstopped neurons in the visual cortex as a substrate for calculating curvature. *Nature, 329*(6138), 438–441. doi:10.1038/329438a0.

Doyon, M. (2011). Husserl and McDowell on the role of concepts in perception. *New Yearbook for Phenomenology and Phenomenological Philosophy, 11*, 42–74.

Dretske, F. (1993). Conscious experience. *Mind, 102*(406), 263–283.

Dretske, F. (2004). Change blindness. *Philosophical Studies, 120*(1–3), 1–18. doi:10.1023/B:PHIL.0000033749.19147.88.

Dretske, F. I. (1995). *Naturalizing the mind.* Cambridge, Mass.: MIT Press.

Drestke, F. (2010). What we see: The texture of conscious experience. In Nanay, *Perceiving the world*, 54–67.

Dreyfus, H. L. (2002). Intelligence without representation. *Phenomenology and the cognitive sciences*, 1(4), 367–383. doi:10.1023/A:1021351606209.

Drummond, J. (2009). Feelings, emotions, and truly perceiving the valuable. *Modern Schoolman*, 86(3), 363–379. doi:10.5840/schoolman2009863/49.

Dubois, D. (2003). Mathematical foundations of discrete and functional systems with strong and weak anticipations. In Butz, Sigaud, and Gérard, *Anticipatory behavior in adaptive learning systems*, 110–32.

Dubois, J., & VanRullen, R. (2011). Visual trails: Do the doors of perception open periodically? *PLoS Biology*, 9(5), e1001056. doi:10.1371/journal.pbio.1001056.

Duhamel, J. R., Colby, C. L., & Goldberg, M. E. (1992). The updating of the representation of visual space in parietal cortex by intended eye movements. *Science*, 255(5040), 90–92.

Dummett, M. A. E. (1996). *Origins of analytical philosophy.* Cambridge. Mass.: Harvard University Press.

Egan, F., & Matthews, R. J. (2006). Doing cognitive neuroscience: A third way. *Synthese*, 153(3), 377–391. doi:10.2307/27653432.

Eliasmith, C., & Anderson, C. H. (2003). *Neural engineering: Computation, representation, and dynamics in neurobiological systems.* Cambridge, Mass.: MIT Press.

Engel, A. K., & Singer, W. (2001). Temporal binding and the neural correlates of sensory awareness. *Trends in Cognitive Sciences*, 5(1), 16–25.

Farah, M. J. (1990/2004). *Visual agnosia* (2nd ed.). Cambridge, Mass.: MIT Press.

Feldman, H., & Friston, K. J. (2010). Attention, uncertainty, and free-energy. *Frontiers in Human Neuroscience*, 4, 215. doi:10.3389/fnhum.2010.00215.

Felleman, D. J., and van Essen, D C. (1991). Distributed hierarchical processing in the primate cerebral cortex. *Cerebral Cortex*, 1(1), 1–47.

Fessard, A. (1936). *Recherches sur l'activité rythmique des nerfs isolés.* Paris: Hermann.

Festinger, L., White, C. W., & Allyn, M. R. (1968). Eye movements and decrements in the Müller-Lyer illusions. *Perception & Psychophysics*, 3, 376–382.

Findlay, J. M., & Gilchrist, I. D. (2003). *Active vision: The psychology of looking and seeing.* Oxford psychology series 37. Oxford: Oxford University Press.

References

Fisette, D. (Ed.). (1999). *Consciousness and intentionality: Models and modalities of attribution*. The Western Ontario series in philosophy of science (Vol. 62). Boston: Kluwer.

Fishman, M. C., & Michael, P. (1973). Integration of auditory information in the cat's visual cortex. *Vision Research, 13*(8), 1415–1419.

Fodor, J. A. (1975). *The language of thought*. Cambridge, Mass.: Harvard University Press.

Fodor, J. A. (1981). *Representations: Philosophical essays on the foundations of cognitive science*. Cambridge, Mass.: MIT Press.

Fodor, J. A. (1983). *Modularity of mind: An essay on faculty psychology*. Cambridge, Mass.: MIT Press.

Fodor, J. A. (1987). *Psychosemantics: The problem of meaning in the philosophy of mind*. Cambridge, Mass.: MIT Press.

Fodor, J., & Pylyshyn, Z. (1981). How direct is visual perception? Some reflections on Gibson's "ecological approach." *Cognition, 9*(2), 139–196.

Freeman, J., & Simoncelli, E. P. (2011). Metamers of the ventral stream. *Nature Neuroscience, 14*(9), 1195–1201. doi:10.1038/nn.2889.

Freeman, W. J. (2000). *How brains make up their minds*. London: Phoenix.

Friston, K. (2012). A free energy principle for biological systems. *Entropy, 14*(11), 2100–2121. doi:10.3390/e14112100.

Friston, K. J., Harrison, L., & Penny, W. (2003). Dynamic causal modelling. *NeuroImage, 19*(4), 1273–1302.

Friston, K. (2005). A theory of cortical responses. *Philosophical Transactions of the Royal Society of London. Series B, Biological Sciences, 360*(1456), 815–836. doi:10.1098/rstb.2005.1622.

Friston, K. (2008). Hierarchical models in the brain. *PLoS Computational Biology, 4*(11), e1000211. doi:10.1371/journal.pcbi.1000211.

Friston, K. (2010). The free-energy principle: A unified brain theory? *Nature Reviews. Neuroscience, 11*(2), 127–138. doi:10.1038/nrn2787.

Friston, Karl. (2013). Life as we know it. *Interface, Journal of the Royal Society 10*(86), 20130475. doi:10.1098/rsif.2013.0475.

Friston, K. J., Daunizeau, J., & Kiebel, S. J. (2009). Reinforcement learning or active inference? *PLoS One, 4*(7), e6421. doi:10.1371/journal.pone.0006421.

Friston, K., & Frith, C. (2015). A duet for one. *Consciousness and Cognition, 36*, 390–405. doi:10.1016/j.concog.2014.12.003.

Friston, K., & Kiebel, S. (2009). Cortical circuits for perceptual inference. *Neural Networks* 22(8), 1093–1104. doi:10.1016/j.neunet.2009.07.023.

Friston, K., Thornton, C., & Clark, A. (2012). Free-energy minimization and the dark-room problem. *Frontiers in Psychology, 3*, 130. doi:10.3389/fpsyg.2012.00130.

Frith, C. D. (2008). Social cognition. *Philosophical Transactions of the Royal Society of London. Series B, Biological Sciences, 363*(1499), 2033–2039. doi:10.1098/rstb.2008.0005.

Frith, C. D. (1992). *The cognitive neuropsychology of schizophrenia. Essays in cognitive psychology.* Hove, UK: Lawrence Erlbaum.

Gallagher, S. (2003). Phenomenology and experimental design: Toward a phenomenologically enlightened experimental science. *Journal of Consciousness Studies, 10*(9–10), 85–99.

Gallagher, S. (2003). Sync-ing in the stream of experience. http://www.theassc.org/files/assc/2570.pdf

Gallagher, S. (2005). *How the body shapes the mind.* Oxford: Clarendon Press.

Gallagher, S., & Zahavi, D. (Eds.). (2007). Special issue on heterophenomenology. *Phenomenology and the Cognitive Sciences, 6*(1–2).

Gallese, V. (2007). The "conscious" dorsal stream: Embodied simulation and its role in space and action awareness. *Psyche, 13*(1), 1–20.

Gangopadhyay, N., Madary, M., & Spicer, F. (Eds.). (2010). *Perception, action, and consciousness: Sensorimotor dynamics and two visual systems.* Oxford: Oxford University Press.

Gärdenfors, P. (2004). *Conceptual spaces: The geometry of thought.* Cambridge, Mass.: MIT Press.

Garrido, M. I., Dolan, R. J., & Sahani, M. (2011). Surprise leads to noisier perceptual decisions. *Perception, 2*(2), 112–120. doi:10.1068/i0411.

Garrido, M. I., Kilner, J. M., Stephan, K. E., & Friston, K. J. (2009). The mismatch negativity: A review of underlying mechanisms. *Clinical Neurophysiology 120*(3), 453–463. doi:10.1016/j.clinph.2008.11.029.

Gazzaniga, M. S. (Ed.). (2009). *The cognitive neurosciences* (4th ed.). Cambridge, Mass.: MIT Press.

Gendler, T., & Hawthorne, J. (Eds.). (2006). *Perceptual experience.* Oxford: Oxford University Press.

Gennaro, R. J. (2004). *Higher-order theories of consciousness.* Amsterdam: John Benjamins Publishing.

References

Gersztenkorn, D., & Lee, A. G. (2015). Palinopsia revamped: A systematic review of the literature. *Survey of Ophthalmology, 60*(1), 1–35. doi:10.1016/j.survophthal.2014.06.003.

Gibson, J. J. (1950). *The perception of the visual world*. Boston: Houghton Mifflin.

Gibson, J. J. (1979). *The ecological approach to visual perception*. Boston, Mass.: Houghton Mifflin.

Goldstein, E. B. (2002). *Sensation and perception* (6th ed.). Belmont, CA: Wadsworth-Cengage Learning.

Gonzalez, C. L. R., Ganel, T., Whitwell, R. L., Morrissey, B., & Goodale, M. A. (2008). Practice makes perfect, but only with the right hand: Sensitivity to perceptual illusions with awkward grasps decreases with practice in the right but not the left hand. *Neuropsychologia, 46*(2), 624–631. doi:10.1016/j.neuropsychologia.2007.09.006.

Gonzalez, C. L. R., Ganel, T., & Goodale, M. A. (2006). Hemispheric specialization for the visual control of action is independent of handedness. *Journal of Neurophysiology 95*(6), 3496–3501. doi:10.1152/jn.01187.2005.

Goodale, M. A., Jakobson, L. S., & Keillor, J. M. (1994). Differences in the visual control of pantomimed and natural grasping movements. *Neuropsychologia, 32*(10), 1159–1178. doi:10.1016/0028-3932(94)90100-7.

Goodale, M. A., Meenan, J. P., Bülthoff, H. H., Nicolle, D. A., Murphy, K. J., & Racicot, C. I. (1994). Separate neural pathways for the visual analysis of object shape in perception and prehension. *Current Biology, 4*(7), 604–610. doi:10.1016/S0960-9822(00)00132-9.

Goodale, M. A., & Milner, A. D. (2004). *Sight unseen: An exploration of conscious and unconscious vision*. Oxford: Oxford University Press.

Gregory, R. (1980). Perceptions and hypotheses. *Philosophical Transactions of the Royal Society of London. Series B, Biological Sciences, 290*(1038), 181–197.

Gregory, R. (1997). Knowledge in perception and illusion. *Philosophical Transactions of the Royal Society of London. Series B, Biological Sciences, 352*(1358), 1121–1127.

Grossberg, S., Mingolla, E., & Ross, W. D. (1997). Visual brain and visual perception: How does the cortex do perceptual grouping? *Trends in Neurosciences, 20*(3), 106–111.

Grush, R. (2004). The emulation theory of representation: Motor control, imagery, and perception. *Behavioral and Brain Sciences, 27*(3), 377–396, discussion 396–442.

Haarmeier, T., Thier, P., Repnow, M., & Petersen, D. (1997). False perception of motion in a patient who cannot compensate for eye movements. *Nature, 389*(6653), 849–852. doi:10.1038/39872.

Hadamard, J. (1898). Les surfaces à courbures opposées et leurs lignes géodésiques. *Journal de Mathématiques Pures et Appliquées*, 5(4), 27–73. http://sites.mathdoc.fr/JMPA/PDF/JMPA_1898_5_4_A3_0.pdf

Haenny, P. E., Maunsell, J. H., & Schiller, P. H. (1988). State dependent activity in monkey visual cortex. II. Retinal and extraretinal factors in V4. *Experimental Brain Research*, 69(2), 245–259.

Hagmann, P., Cammoun, L., Gigandet, X., Meuli, R., Honey, C. J., van Wedeen, J., et al. (2008). Mapping the structural core of human cerebral cortex. *PLoS Biology*, 6(7), e159. doi:10.1371/journal.pbio.0060159.

Hahn, M., & Ramberg, B. T. (Eds.). (2003). *Reflections and replies: Essays on the philosophy of Tyler Burge*. Cambridge, Mass.: MIT Press.

Hamzei, F., Rijntjes, M., Dettmers, C., Glauche, V., Weiller, C., & Büchel, C. (2003). The human action recognition system and its relationship to Broca's area: An fMRI study. *NeuroImage*, 19(3), 637–644.

Han, F., Caporale, N., & Dan, Y. (2008). Reverberation of recent visual experience in spontaneous cortical waves. *Neuron*, 60(2), 321–327. doi:10.1016/j.neuron.2008.08.026.

Hansen, T., Olkkonen, M., Walter, S., & Gegenfurtner, K. R. (2006). Memory modulates color appearance. *Nature Neuroscience*, 9(11), 1367–1368. doi:10.1038/nn1794.

Hansen, T., Pracejus, L., & Gegenfurtner, K. R. (2009). Color perception in the intermediate periphery of the visual field. *Journal of Vision* 9(4), 26.1–12. doi:10.1167/9.4.26.

Hardin, C. L. (1988). *Color for philosophers: Unweaving the rainbow*. Indianapolis: Hackett.

Harman, G. (1990). The intrinsic quality of experience. *Philosophical Perspectives*, 4, 31–52.

Hayhoe, M., & Ballard, D. (2005). Eye movements in natural behavior. *Trends in Cognitive Sciences*, 9(4), 188–194. doi:10.1016/j.tics.2005.02.009.

Haynie, D. T. (2008). *Biological thermodynamics* (2nd ed.). Cambridge, UK: Cambridge University Press.

Held, R., & Hein, A. (1963). Movement produced stimulation in the development of visually guided behavior. *Journal of Comparative and Physiological Psychology*, 56, 873–876.

Helmholtz, H. (1867). *Handbuch der Physiologischen Optik*. Leipzig: Leopold Voss.

Hess, R. H., Baker, C. L., & Zihl, J. (1989). The "motion-blind" patient: Low-level spatial and temporal filters. *Journal of Neuroscience*, 9(5), 1628–1640.

References

Hettinger, L. J., & Riccio, G. E. (1992). Visually induced motion sickness in virtual environments. *Presence*, *1*(3), 306–310. doi:10.1162/pres.1992.1.3.306.

Heyes, C. (2011). Automatic imitation. *Psychological Bulletin*, *137*(3), 463–483. doi:10.1037/a0022288.

Hinton, G. (2006). To recognize shapes, first learn to generate images. http://www.cs.toronto.edu/~fritz/absps/montrealTR.pdf

Hinton, G., Dayan, P., Frey, B., & Neal, R. (1995). The "wake-sleep" algorithm for unsupervised neural networks. *Science*, *268*(5214), 1158–1161. doi:10.1126/science.7761831.

Hinton, G. E. (2007). Learning multiple layers of representation. *Trends in Cognitive Sciences*, *11*(10), 428–434. doi:10.1016/j.tics.2007.09.004.

Hinton, J. M. (1973). *Experiences: An inquiry into some ambiguities*. Clarendon library of logic and philosophy. Oxford: Clarendon Press.

Hinton, J. M. (1967). Visual experiences. *Mind*, *LXXVI*(302), 217–227. doi:10.1093/mind/LXXVI.302.217.

Hodgkin, A. L. (1948). The local electric changes associated with repetitive action in a non-medullated axon. *Journal of Physiology*, *107*(2), 165–181. doi:10.1113/jphysiol.1948.sp004260.

Hohwy, J. (2013). *The predictive mind*. Oxford: Oxford University Press.

Hohwy, J. (2012). Attention and conscious perception in the hypothesis testing brain. *Frontiers in Psychology*, *3*, 96. doi:10.3389/fpsyg.2012.00096.

Hopp, W. (2008). Husserl on sensation, perception, and interpretation. *Canadian Journal of Philosophy*, *38*(2), 219–245. doi:10.1353/cjp.0.0013.

Hopp, W. (2010). How to think about nonconceptual content. *New Yearbook for Phenomenology and Phenomenological Philosophy*, *10*(1), 1–24.

Hopp, W. (2013). No such look: Problems with the dual content theory. *Phenomenology and the Cognitive Sciences*, *12*(4), 813–833.

Horn, G., & Hill, R. M. (1969). Modifications of receptive fields of cells in the visual cortex occurring spontaneously and associated with bodily tilt. *Nature*, *221*(5176), 186–188.

Horton, J. C., & Trobe, J. D. (1999). Akinetopsia from nefazodone toxicity. *American Journal of Ophthalmology*, *128*(4), 530–531.

Hubel, D. H., & Wiesel, T. N. (1959). Receptive fields of single neurones in the cat's striate cortex. *Journal of Physiology*, *148*, 574–591.

References

Hume, D., & Steinberg, E. (1993). *An enquiry concerning human understanding; [with] A letter from a gentleman to his friend in Edinburgh; [and] An abstract of a Treatise of human nature*. (2nd ed.) Indianapolis: Hackett.

Hurley, S. L. (1998). *Consciousness in action*. Cambridge, Mass.: Harvard University Press.

Hurley, S. L. (2010). The varieties of externalism. In Menary, *The extended mind*, 101–154.

Husserl, E. (1993). *Logische Untersuchungen* (3 Vols.). Tübingen: Max Niemeyer. (1900).

Husserl, E. (1965). Philosophy as rigorous science. In Lauer, *Phenomenology and the crisis of philosophy*, 71–147.

Husserl, E. 1980. *Ideas pertaining to a pure phenomenology and to a phenomenological philosophy*. Edmund Husserl collected works (Vols. 1–2). The Hague: Kluwer.

Husserl, E. (2001). *Analyses concerning passive and active synthesis: Lectures on transcendental logic*. Edmund Husserl collected works (Vol. 9). Dordrecht: Kluwer.

Husserl, E., & Biemel, W. (1968). *Phänomenologische Psychologie. Husserliana 9*. The Hague: Martinus Nijhoff.

Husserl, E., & Biemel, W. (1976). *Die Krisis der europäischen Wissenschaften und die transzendentale Phänomenologie: Eine Einleitung in die Phänomenologische Philosophie*. 2. Aufl, photomechanischer Nachdr. The Hague: Martinus Nijhoff.

Husserl, E., & Boehm, R. (1966). *Zur Phänomenologie des inneren Zeitbewusstseins (1893–1917)*. Husserliana: Gesammelte werke (Vol. 10). The Hague: Martinus Nijhoff.

Husserl, E., van Breda, H. L., IJsseling, S., Schuhmann, K., Bernet, R., Eley, L., et al. (1976). *Husserliana Ideen zu einter reinen Phänomenologie und phänomenologischen Philosophie: Gesammelte Werke. Text der 1-3*. Dordrecht: Kluwer.

Husserl, E., & Claesges, U. (1973b). *Ding und Raum. Vorlesungen 1907*. Husserliana: Gesammelte werke bd. 16 . The Hague: Martinus Nijhoff.

Husserl, E., & Fleischer, M. (1966). *Analysen zur passiven Synthesis: Aus Vorlesungs- und Forschungsmanuskripten 1918–1926*. Husserliana: Gesammelte werke bd. 11. The Hague: Martinus Nijhoff.

Husserl, E., & Kern, I. (1973a). *Zur Phanomenologie der intersubjektivitat: Texte Aus dem nachlass* (Vols. 13–15). The Hague: Martinus Nijhoff.

Husserl, E., & Landgrebe, L. (1973c). *Experience and judgment: Investigations in a genealogy of logic*. Rev. and ed. by L. Landgrebe. Northwestern University studies in phenomenology and existential philosophy. Evanston: Northwestern University Press.

References

Husserl, E., & Landgrebe, L. (1985). *Erfahrung und Urteil: Untersuchungen zur Genealogie der Logik. 6., verb. Aufl. mit Kolumnentiteln. Philosophische Bibliothek* (Vol. 280). Hamburg: F. Meiner Verlag.

Husserl, E., & Rojcewicz, R. (1997). *Thing and space. Collected works.* Dordrecht: Kluwer.

Husserl, E., Rojcewicz, R., & Schuwer, A. (1989). *Studies in the phenomenology of constitution.* Edmund Husserl collected works (Vol. 3). Dordrecht: Kluwer.

Husserl, E., Schuhmann, K., & Biemel, M. (1969). *Ideen zu einer reinen Phänomenologie und phänomenologischen Philosophie.* The Hague: Martinus Nijhoff.

Hutto, D. D. (2012). *Folk Psychological Narratives: The Sociocultural Basis of Understanding Reasons.* MIT Press.

Hutto, D. D., & Myin, E. (2013). *Radicalizing enactivism: Basic minds without content.* Cambridge, Mass.: MIT Press.

Ingarden, R. (1975). *On the motives which led Husserl to transcendental idealism. Phaenomenologica 64.* The Hague: Martinus Nijhoff.

Izhikevich, E. M. (2010). *Dynamical systems in neuroscience: The geometry of excitability and bursting.* Cambridge, Mass.: MIT Press.

Jackendoff, R. (1989). *Consciousness and the computational mind.* Explorations in cognitive science. Cambridge, Mass.: MIT Press.

Jackson, F. (1977). *Perception: A representative theory.* Cambridge, UK: Cambridge University Press.

Jacob, P., & Jeannerod, M. (2003). *Ways of seeing: The scope and limits of visual cognition.* Oxford cognitive science series. New York: Oxford University Press.

James, W. 1950. *The principles of psychology.* New York: Dover. (1918).

Jiang, Y., Zhou, K., & He, S. (2007). Human visual cortex responds to invisible chromatic flicker. *Nature Neuroscience, 10*(5), 657–662. doi:10.1038/nn1879.

Johansson, G. (1973). Visual perception of biological motion and a model for its analysis. *Perception & Psychophysics, 14*(2), 201–211. doi:10.3758/bf03212378.

Johnson, M. H., Dziurawiec, S., Ellis, H., & Morton, J. (1991). Newborns' preferential tracking of face-like stimuli and its subsequent decline. *Cognition, 40*(1–2), 1–19. doi:10.1016/0010-0277(91)90045-6.

Juola, P. & Plunkett, K. (2000). Why double dissociations don't mean much. In Cohen, Johnston, & Plunkett, *Exploring cognition,* 319–327.

Kaiser, M. D., & Shiffrar, M. (2009). The visual perception of motion by observers with autism spectrum disorders: A review and synthesis. *Psychonomic Bulletin & Review, 16*(5), 761–777. doi:10.3758/PBR.16.5.761.

Kanizsa, G. (1969). Perception, past experience, and the impossible experiment. *Acta Psychologica, 59*, 66–96.

Kanizsa, G. (1976). Subjective contours. *Scientific American, 234*, 48–52.

Kanizsa, G. (1979). *Organization in vision: Essays on gestalt perception*. New York: Praeger.

Kanizsa, G. (1985). Seeing and thinking. *Acta Psychologica, 59*, 23–33.

Kanizsa, G., & Gerbino, W. (1982). Amodal completion: Seeing or thinking? In Beck, *Organization and representation in perception*, 167–90.

Kant, I. (1998). *Kritik der reinen Vernunft*. Hamburg: Meiner. (1781).

Kelly, S. D. (2001). Demonstrative concepts and experience. *Philosophical Review, 110*(3), 397–420.

Kelly, S. D. (2004). Seeing things in Merleau-Ponty. In Carman & Hansen, *The Cambridge companion to Merleau-Ponty*, 74–110.

Kelly, S. D. (2005). Temporal awareness. In Smith & Thomasson, *Phenomenology and philosophy of mind*, 222–234).

Kenet, T., Bibitchkov, D., Tsodyks, M., Grinvald, A., & Arieli, A. (2003). Spontaneously emerging cortical representations of visual attributes. *Nature, 425*(6961), 954–956. doi:10.1038/nature02078.

Klein, C. (2013). Review of Robert Shulman's Brain imaging: What it can (and cannot) tell us about consciousness. http://ndpr.nd.edu/news/40439-brain-imaging-what-it-can-and-cannot-tell-us-about-consciousness/.

Koch, C., & Davis, J. L. (Eds.). (1994). *Large-scale neuronal theories of the brain*. Cambridge, Mass.: MIT Press.

Kohler, I. (1964). *The formation and transformation of the perceptual world*. Psychological issues (Vol. 3, No. 4, monograph 12). New York: International Universities Press.

Konen, C. S., & Kastner, S. (2008). Two hierarchically organized neural systems for object information in human visual cortex. *Nature Neuroscience, 11*(2), 224–231. doi:10.1038/nn2036.

Konvalinka, I., Xygalatas, D., Bulbulia, J., Schjødt, U., Jegindø, E.-M., Wallot, S., et al. (2011). Synchronized arousal between performers and related spectators in a fire-walking ritual. *Proceedings of the National Academy of Sciences of the United States of America, 108*(20), 8514–8519. doi:10.1073/pnas.1016955108.

References

Kozlowski, L. T., & Cutting, J. E. (1977). Recognizing the sex of a walker from a dynamic point-light display. *Perception & Psychophysics, 21*(6), 575–580. doi:10.3758/BF03198740.

Króliczak, G., Heard, P., Goodale, M. A., & Gregory, R. L. (2006). Dissociation of perception and action unmasked by the hollow-face illusion. *Brain Research, 1080*(1), 9–16. doi:10.1016/j.brainres.2005.01.107.

Kühn, S., Haggard, P., & Brass, M. (2009). Intentional inhibition: How the "veto-area" exerts control. *Human Brain Mapping, 30*(9), 2834–2843. doi:10.1002/hbm.20711.

Kveraga, K., Ghuman, A. S., & Bar, M. (2007). Top-down predictions in the cognitive brain. *Brain and Cognition, 65*(2), 145–168. doi:10.1016/j.bandc.2007.06.007.

Laasik, K. (2014). Constitutive strata and the dorsal stream. *Phenomenology and the Cognitive Sciences, 13*(3), 419–435. doi:10.1007/s11097-013-9306-2.

Land, M., Mennie, N., & Rusted, J. (1999). The roles of vision and eye movements in the control of activities of daily living. *Perception, 28*(11), 1311–1328. doi:10.1068/p2935.

Lashley, K. (Ed.). (1934). *Instinctive behavior*. New York: International Universities Press.

Lauer, Q. (Ed.). (1965). *Phenomenology and the crisis of philosophy*. New York: Harper. (1910).

Lawson, R., Bertamini, M., & Liu, D. (2007). Overestimation of the projected size of objects on the surface of mirrors and windows. *Journal of Experimental Psychology: Human Perception and Performance, 33*(5), 1027–1044. doi:10.1037/0096-1523.33.5.1027.

Lê, S., Cardebat, D., Boulanouar, K., Hénaff, M.-A., Michel, F., Milner, D., et al. (2002). Seeing, since childhood, without ventral stream: A behavioural study. *Brain, 125*(1), 58–74. doi:10.1093/brain/awf004.

Lee, L., Friston, K., & Horwitz, B. (2006). Large-scale neural models and dynamic causal modelling. *NeuroImage, 30*(4), 1243–1254. doi:10.1016/j.neuroimage.2005.11.007.

Lee, T. S., & Mumford, D. (2003). Hierarchical Bayesian inference in the visual cortex. *Journal of the Optical Society of America. A, Optics, Image Science, and Vision, 20*(7), 1434–1448.

Leopold, D. A., Murayama, Y., & Logothetis, N. K. (2003). Very slow activity fluctuations in monkey visual cortex: implications for functional brain imaging. *Cerebral Cortex, 13*(4), 422–433.

Levin, D. M. (1968). Induction and Husserl's theory of eidetic variation. *Philosophy and Phenomenological Research, 29*(1), 1. doi:10.2307/2105814.

Levin, J. (2013). Functionalism. The Stanford Encyclopedia of Philosophy. http://plato.stanford.edu/archives/fall2013/entries/functionalism/.

Levine, J. (1983). Materialism and qualia: The explanatory gap. *Pacific Philosophical Quarterly, 64*, 354–361.

Lewis, C. I. (1946). *An analysis of knowledge and valuation.* La Salle, Ill.: Open Court.

Lewis, D. (1980). Veridical hallucination and prosthetic vision. *Australasian Journal of Philosophy, 58*, 239–249.

Lewis, D. (1981). Mad pain and Martian pain. In Block, *Readings in philosophy of psychology*, 67–106.

Lhermitte, F., Pillon, B., & Serdaru, M. (1986). Human anatomy and the frontal lobes. Part I: Imitation and utilization behavior. *Annals of Neurology, 19*, 326–334.

Lind, D. A., & Marcus, B. (1995). *An introduction to symbolic dynamics and coding.* Cambridge, UK: Cambridge University Press.

Lindsay, P. H., & Norman, D. A. (1977). *Human information processing: An introduction to psychology* (2nd ed.). New York: Academic Press.

Livingstone, M., & Hubel, D. (1988). Segregation of form, color, movement, and depth: Anatomy, physiology, and perception. *Science, 240*(4853), 740–749. doi:10.1126/science.3283936.

Llinás, R. (1988). The intrinsic electrophysiological properties of mammalian neurons: Insights into central nervous system function. *Science, 242*(4886), 1654–1664. doi:10.1126/science.3059497.

Logothetis, N. K. (2008). What we can do and what we cannot do with fMRI. *Nature, 453*(7197), 869–878. doi:10.1038/nature06976.

Logue, H. (2013). Visual experience of natural kind properties: Is there any fact of the matter? *Philosophical Studies: An International Journal for Philosophy in the Analytic Tradition, 162*(1), 1–12.

Lumer, E. D., & Rees, G. (1999). Covariation of activity in visual and prefrontal cortex associated with subjective visual perception. *Proceedings of the National Academy of Sciences of the United States of America, 96*(4), 1669–1673.

Lutz, A. (2002). Toward a neurophenomenology as an account of generative passages: A first empirical case study. *Phenomenology and the Cognitive Sciences, 1*(2), 133–167. doi:10.1023/A:1020320221083.

References

Lutz, A., Lachaux, J.-P., Martinerie, J., & Varela, F. J. (2002). Guiding the study of brain dynamics by using first-person data: Synchrony patterns correlate with ongoing conscious states during a simple visual task. *Proceedings of the National Academy of Sciences of the United States of America, 99*(3), 1586–1591. doi:10.1073/pnas.032658199.

Lutz, A., & Thompson, E. (2003). Neurophenomenology—integrating subjective experience and brain dynamics in the neuroscience of consciousness. *Journal of Consciousness Studies 10*(9–10), 31–52.

Mac Lane, S., & Moerdijk, I. (1992). *Sheaves in geometry and logic: A first introduction to topos theory*. New York: Springer-Verlag.

Mack, A., & Rock, I. (1998). *Inattentional blindness*. Cambridge, Mass.: MIT Press.

Macknik, S., & Martinez-Conde, S. (2009). The role of feedback in visual attention and awareness. In Gazzaniga, *The cognitive neurosciences*, 1165–1180.

Macpherson, F. (2012). Cognitive penetration of colour experience: Rethinking the issue in light of an indirect mechanism. *Philosophy and Phenomenological Research, 84*(1), 24–62. doi:10.1111/j.1933-1592.2010.00481.x.

Madary, M. (2008). Specular highlights as a guide to perceptual content. *Philosophical Psychology, 21*(5), 629–639. doi:10.1080/09515080802412347.

Madary, M. (2011). The dorsal stream and the visual horizon. *Phenomenology and the Cognitive Sciences, 10*, 423–438.

Madary, M. (2012a). Husserl on perceptual constancy. *European Journal of Philosophy, 20*, 145–165.

Madary, M. (2012b). How would the world look if it looked as if it were encoded as an intertwined set of probability density distributions? *Frontiers in Psychology, 3*, 419. doi:10.3389/fpsyg.2012.00419.

Madary, M. (2012c). Showtime at the Cartesian theater? Vehicle externalism and dynamical explanations. In Paglieri, *Consciousness in interaction*, 59–72.

Madary, M. (2013a). Anticipation and variation in visual content. *Philosophical Studies: An International Journal for Philosophy in the Analytic Tradition, 165*(2), 335–347. doi:10.2307/42920510.

Madary, M. (2013b). Placing area MT in context. *Journal of Consciousness Studies, 20*(5–6), 93–104.

Madary, M. (2014a). Visual experience. In Shapiro, *The Routledge handbook of embodied cognition*, 263–271.

Madary, M. (2014b). Perceptual presence without counterfactual richness. *Cognitive Neuroscience, 5*(2), 131–133. doi:10.1080/17588928.2014.907257.

Madary, M. (2015a). Extending the explanandum for predictive processing. In Metzinger & Windt, *Open mind: Philosophy and the mind sciences in the 21st century*.

Madary, M. (2015b). Varieties of presence. *Philosophical Quarterly*. doi:10.1093/pq/pqv031.

Madary, M., & Metzinger, T. K. (2016). Recommendations for good scientific practice and the consumers of VR-technology. *Frontiers in Robotics and AI*, *3*(Suppl. 3), 235. doi:10.3389/frobt.2016.00003.

Maier, A., Logothetis, N. K., & Leopold, D. A. (2007). Context-dependent perceptual modulation of single neurons in primate visual cortex. *Proceedings of the National Academy of Sciences of the United States of America*, *104*(13), 5620–5625. doi:10.1073/pnas.0608489104.

Manjaly, Z. M., Marshall, J. C., Stephan, K. E., Gurd, J. M., Zilles, K., & Fink, G. R. (2003). In search of the hidden: An fMRI study with implications for the study of patients with autism and with acquired brain injury. *NeuroImage*, *19*(3), 674–683.

Marbach, E. (1984). On using intentionality in empirical phenomenology: The problem of mental images. *Dialectica*, *38*(2–3), 209–229. doi:10.1111/j.1746-8361.1984.tb01245.x.

Marbach, E., & Husserl, E. (1980). *Phantasie, Bildbewusstsein, Erinnerung: Zur Phänomenologie der anschaulichen Vergegenwärtigungen: Texte aus dem Nachlass (1898–1925)*. Husserliana (Vol. 23). The Hague: Martinus Nijhoff.

Markman, A. B., & Dietrich, E. (2000). Extending the classical view of representation. *Trends in Cognitive Sciences*, *4*(12), 470–475. doi:10.1016/S1364-6613(00)01559-X.

Marr, D. (1983/2010). *Vision: A computational investigation into the human representation and processing of visual information*. Cambridge, Mass.: MIT Press.

Martin, M. (2002). The transparency of experience. *Mind & Language*, *17*, 376–389.

Martin, M. (2006). On being alienated. In Gendler & Hawthorne, *Perceptual experience*, 354–410.

Martin, W. (2005). Husserl and the logic of consciousness. In Smith & Thomasson, *Phenomenology and philosophy of mind*, 203–221.

Martinez-Conde, S., Cudeiro, J., Grieve, K. L., Rodriguez, R., Rivadulla, C., & Acuña, C. (1999). Effects of feedback projections from area 18 layers 2/3 to area 17 layers 2/3 in the cat visual cortex. *Journal of Neurophysiology*, *82*(5), 2667–2675.

Martinez-Conde, S. (2009). Microsaccades: A neurophysiological analysis. *Trends in Neurosciences*, *32*, 463–475. doi:10.1016/j.tins.2009.05.006.

References

Masuda, T., & Nisbett, R. E. (2001). Attending holistically versus analytically: Comparing the context sensitivity of Japanese and Americans. *Journal of Personality and Social Psychology, 81*(5), 922–934. doi:10.1037/0022-3514.81.5.922.

Matthen, M. (2010). Two visual systems and the feeling of presence. In Gangopadhyay, Madary, and Spicer, *Perception, action, and consciousness*, 107–24.

Mattler, U. (2003). Priming of mental operations by masked stimuli. *Perception & Psychophysics, 65*(2), 167–187. doi:10.3758/BF03194793.

Maturana, H. R., & Varela, F. J. (1992). *The tree of knowledge: The biological roots of human understanding*. (Rev. ed.) Boston: Shambhala. (1980).

McDermott, K. C., Malkoc, G., Mulligan, J. B., & Webster, M. A. (2010). Adaptation and visual salience. *Journal of Vision, 10*(13), 17. doi:10.1167/10.13.17.

McDowell, J. (1982). Criteria, defeasability and knowledge. *Proceedings of the British Academy, 68*, 455–479.

McDowell, J. (2009). *Having the world in view: Essays on Kant, Hegel, and Sellars*. Cambridge, Mass.: Harvard University Press.

McDowell, J. H. (1994). *Mind and world*. Cambridge, Mass.: Harvard University Press.

McGinn, C. (1989). *Mental content*. Oxford: Basil Blackwell.

Meadows, J. C., & Munro, S. S. (1977). Palinopsia. *Journal of Neurology, Neurosurgery, and Psychiatry, 40*(1), 5–8.

Melcher, D. (2007). Predictive remapping of visual features precedes saccadic eye movements. *Nature Neuroscience, 10*(7), 903 907. doi:10.1038/nn1917.

Melcher, D., & Colby, C. L. (2008). Trans-saccadic perception. *Trends in Cognitive Sciences, 12*(12), 466–473. doi:10.1016/j.tics.2008.09.003.

Menary, R. (Ed.). (2010). *The extended mind*. Cambridge, MA: MIT Press.

Merigan, W. H., & Maunsell, J. H. (1993). How parallel are the primate visual pathways? *Annual Review of Neuroscience, 16*, 369–402. doi:10.1146/annurev.ne.16.030193.002101.

Merleau-Ponty, M. (1945). *Phénoménologie de la perception. Bibliothèque des idées*. Paris: Gallimard.

Merleau-Ponty, M. (1962). *Phenomenology of perception*. Trans. C. Smith. International Library of Philosophy and Scientific Method. London: Routledge & Kegan Paul.

Metzinger, T. (Ed.). (2000). *Neural correlates of consciousness: Empirical and conceptual questions*. Cambridge, Mass.: MIT Press.

Metzinger, T. (Ed.). (2003). *Being no one: The self-model theory of subjectivity*. Cambridge, Mass.: MIT Press.

Metzinger, T. (Ed.). (2009). *The ego tunnel: The science of the mind and the myth of the self*. New York: Basic Books.

Metzinger, T. & Windt, J. M. (Eds.), *Open mind: Philosophy and the mind sciences in the 21st century*. Frankfurt am Main: MIND Group. www.open-mind.net.

Milner, A. D., & Melvyn, A. G. (1995). *The visual brain in action*. Oxford science publications no. 27. Oxford: Oxford University Press.

Milner, D. and Goodale, M. (2010). Cortical visual systems for perception and action. In Gangopadhyay, Madary, and Spicer, *Perception, action, and consciousness*, 71–94.

Mole, C. (2011). *Attention is cognitive unison: An essay in philosophical psychology*. Oxford: Oxford University Press.

Montague, M. (2007). Against propositionalism. *Noûs, 41*(3), 503–518. doi:10.1111/j.1468-0068.2007.00657.x.

Moore, G. E. (1953). *Some main problems of philosophy*. London: George Allen & Unwin.

Moore, T., & Armstrong, K. M. (2003). Selective gating of visual signals by microstimulation of frontal cortex. *Nature, 421*(6921), 370–373. doi:10.1038/nature01341.

Moore, T., Armstrong, K. M., & Fallah, M. (2003). Visuomotor origins of covert spatial attention. *Neuron, 40*(4), 671–683. doi:10.1016/S0896-6273(03)00716-5.

Morrell, F. (1972). Visual system's view of acoustic space. *Nature, 238*, 44–46.

Morse, M., & Hedlund, G. A. (1938). Symbolic dynamics. *American Journal of Mathematics, 60*(4), 815. doi:10.2307/2371264.

Mulder, H. L., & van de Velde-Schlick, B. (Eds.). (1979). *Philosophical papers*. Dordrecht: Reidel.

Mulligan, K. (1995). Perception. In Smith & Smith, *The Cambridge companion to Husserl*, 168–238.

Mulligan, K. (1999). Perception, predicates, and particulars. In Fisette, *Consciousness and intentionality*, 163–94.

Murray, S. O., Boyaci, H., & Kersten, D. (2006). The representation of perceived angular size in human primary visual cortex. *Nature Neuroscience, 9*(3), 429–434. doi:10.1038/nn1641.

References

Näätänen, R., Gaillard, A. W. K., & Mäntysalo, S. (1978). Early selective-attention effect on evoked potential reinterpreted. *Acta Psychologica, 42*(4), 313–329. doi:10.1016/0001-6918(78)90006-9.

Nagel, T. (1974). What is it like to be a bat? *Philosophical Review, 83*, 435–450.

Nanay, B. (2009). How speckled is the hen? *Analysis, 69*(3), 499–502.

Nanay, B. (Ed.). (2010). *Perceiving the world*. Oxford: Oxford University Press.

Nanay, B. (Ed.). (2012). Bayes or determinables? What does the bidirectional hierarchical model of brain functions tell us about the nature of perceptual representation? *Frontiers in Psychology, 3*, 500. doi:10.3389/fpsyg.2012.00500.

Nassi, J. J., & Callaway, E. M. (2006). Multiple circuits relaying primate parallel visual pathways to the middle temporal area. *Journal of Neuroscience, 26*(49), 12789–12798. doi:10.1523/JNEUROSCI.4044-06.2006.

Nassi, J. J., & Callaway, E. M. (2009). Parallel processing strategies of the primate visual system. *Nature Reviews. Neuroscience, 10*(5), 360–372. doi:10.1038/nrn2619.

Nawrot, M. P., Boucsein, C., Molina, V. R., Riehle, A., Aertsen, A., & Rotter, S. (2008). Measurement of variability dynamics in cortical spike trains. *Journal of Neuroscience Methods, 169*(2), 374–390. doi:10.1016/j.jneumeth.2007.10.013.

Newton, J. R., & Eskew, R. T. (2003). Chromatic detection and discrimination in the periphery: A postreceptoral loss of color sensitivity. *Visual Neuroscience, 20*(5), 511–521.

Nichols, S., & Stich, S. P. (2003). *Mindreading: An integrated account of pretence, self-awareness, and understanding other minds*. Oxford cognitive science series. Oxford: Oxford University Press.

Nisbett, R., & Wilson, T. (1977). Telling more than we can know. *Psychological Review, 84*, 231–259.

Noë, A. (2002). Is the visual world a grand illusion? *Journal of Consciousness Studies 9*(5–6).

Noë, A. (2004). *Action in perception*. Representation and mind. Cambridge, Mass.: MIT Press.

Noë, A. (2010). Vision without representation. In Gangopadhyay, Madary, and Spicer, *Perception, action, and consciousness*, 245–256.

Noë, A. (2012). *Varieties of presence*. Cambridge, Mass.: Harvard University Press.

Noë, A., & Thompson, E. (Eds.). (2002). *Vision and mind: Selected readings in the philosophy of perception*. Cambridge, Mass.: MIT Press.

Noë, A., & Thompson, E. (2004). Are there neural correlates of consciousness? *Journal of Consciousness Studies*, *11*(1), 3–28.

Norton, J. W., & Corbett, J. J. (2000). Visual perceptual abnormalities: Hallucinations and illusions. *Seminars in Neurology*, *20*(1), 111–121.

Nowak, L. G., & Bullier, J. (1997). The timing of information transfer in the visual system. In Rockland et al., *Cerebral cortex: Extrastriate cortex in humans*, 205–233.

O'Regan, J. K., & Block, N. (2012). Discussion of J. Kevin O'Regan's Why red doesn't sound like a bell: Understanding the feel of consciousness. *Review of Philosophy and Psychology*, *3*(1), 89–108. doi:10.1007/s13164-012-0090-7.

Öğmen, H. (1993). A neural theory of retino-cortical dynamics. *Neural Networks*, *6*(2), 245–273. doi:10.1016/0893-6080(93)90020-W.

Öğmen, H., & Breitmeyer, B. G. (Eds.). (2006). *The first half second: The microgenesis and temporal dynamics of unconscious and conscious visual processes*. Cambridge, Mass.: MIT Press.

Öğmen, H., Breitmeyer, B., Bedell, H. (2006). Dynamics of perceptual epochs probed by dissociation phenomena in masking. In Öğmen and Breitmeyer, *The first half second*, 149–170.

Olivers, C. N. L., & Nieuwenhuis, S. (2005). The beneficial effect of concurrent task-irrelevant mental activity on temporal attention. *Psychological Science* *16*(4), 265–269. doi:10.1111/j.0956-7976.2005.01526.x.

O'Regan, J. K. (1990). Eye movements and reading. *Reviews of Oculomotor Research*, *4*, 395–453.

O'Regan, J. K. (2011). *Why red doesn't sound like a bell: Understanding the feel of consciousness*. New York: Oxford University Press.

O'Regan, J. K., & Noë, A. (2001). A sensorimotor account of vision and visual consciousness. *Behavioral and Brain Sciences*, *24*(5), 939–973, discussion 973–1031.

Overgaard, M., Rote, J., Mouridsen, K., & Ramsøy, T. Z. (2006). Is conscious perception gradual or dichotomous? A comparison of report methodologies during a visual task. *Consciousness and Cognition*, *15*(4), 700–708. doi:10.1016/j.concog.2006.04.002.

Overgaard, S. (2010). On the looks of things. *Pacific Philosophical Quarterly*, *91*(2), 260–284.

Paglieri, F. (Ed.). (2012). *Consciousness in interaction*. Amsterdam: John Benjamins Publishing.

Palmer, S. E. (1999). *Vision science: Photons to phenomenology*. Cambridge, Mass.: MIT Press.

References

Pappas, J. M., Fishel, S. R., Moss, J. D., Hicks, J. M., & Leech, T. D. (2005). An eye-tracking approach to inattentional blindness. *Proceedings of the Human Factors and Ergonomics Society Annual Meeting, 49*(17), 1658–1662. doi:10.1177/154193120504901734.

Parga, N., & Abbott, L. F. (2007). Network model of spontaneous activity exhibiting synchronous transitions between up and down states. *Frontiers in Neuroscience, 1*(1), 57–66. doi:10.3389/neuro.01.1.1.004.2007.

de Pasquale, R., & Sherman, S. M. (2011). Synaptic properties of corticocortical connections between the primary and secondary visual cortical areas in the mouse. *Journal of Neuroscience, 31*(46), 16494–16506. doi:10.1523/JNEUROSCI.3664-11.2011.

Peacocke, C. (1983). *Sense and content: Experience, thought, and their relations*. Oxford: Oxford University Press.

Peacocke, C. (1992). *A study of concepts*. Cambridge, Mass.: MIT Press.

Pelak, V. S., & Hoyt, W. F. (2009). Symptoms of akinetopsia associated with traumatic brain injury and Alzheimer's disease. *Neuro-Ophthalmology (Aeolus Press), 29*(4), 137–142. doi:10.1080/01658100500218046.

Peterhans, E., & von der Heydt, R. (1991). Subjective contours—bridging the gap between psychophysics and physiology. *Trends in Neurosciences, 14*(3), 112–119.

Peters, A., B. R. Payne, and J. Budd. (1994). A numerical analysis of the geniculocortical input to striate cortex in the monkey. *Cerebral Cortex, 4*(3), 215–229.

Petersen, C. C. H., Hahn, T. T. G., Mehta, M., Grinvald, A., and Sakmann, B. (2003). Interaction of sensory responses with spontaneous depolarization in layer 2/3 barrel cortex. *Proceedings of the National Academy of Sciences of the United States of America 100*(23), 13638–13643. doi:10.1073/pnas.2235811100.

Petitot, J. (1999). Morphological eidetics for a phenomenology of perception. In Petitot, *Naturalizing phenomenology*, 330–371.

Petitot, J. (2010). Sheaf Mereology and Space Cognition. In Carsetti, *Functional models of cognition*, 49–74.

Petitot, J., Varela, F., Pachoud, B., Roy, J-M. (Eds.). (1999). *Naturalizing phenomenology: Issues in contemporary phenomenology and cognitive science*. Stanford, Calif.: Stanford University Press.

Phillips, I. (2010). Perceiving temporal properties. *European Journal of Philosophy, 18*(2), 176–202. doi:10.1111/j.1468-0378.2008.00299.x.

Phillips, I. (2011). Perception and iconic memory: What Sperling doesn't show. *Mind & Language, 26*(4), 381–411.

Pinel, J. (2003). *Biopsychology*. (5th ed.) Boston, Mass.: Allyn and Bacon.

Pisella, L., Binkofski, F., Lasek, K., Toni, I., & Rossetti, Y. (2006). No double-dissociation between optic ataxia and visual agnosia: Multiple sub-streams for multiple visuo-manual integrations. *Neuropsychologia, 44*(13), 2734–2748. doi:10.1016/j.neuropsychologia.2006.03.027.

Plunkett, K., & Bandelow, S. (2006). Stochastic approaches to understanding dissociations in inflectional morphology. *Brain and Language, 98*(2), 194–209. doi:10.1016/j.bandl.2006.04.014.

Port, R. F., & van Gelder, T. (1995). *Mind as motion: Explorations in the dynamics of cognition*. Cambridge, Mass.: MIT Press.

Posner, M. I., Snyder, C. R., & Davidson, B. J. (1980). Attention and the detection of signals. *Journal of Experimental Psychology, 109*(2), 160–174.

Prime, S. L., Vesia, M., & Crawford, J. D. (2011). Cortical mechanisms for trans-saccadic memory and integration of multiple object features. *Philosophical Transactions of the Royal Society of London. Series B, Biological Sciences, 366*(1564), 540–553. doi:10.1098/rstb.2010.0184.

Prinz, J. (2006a). Is the mind really modular? In Stainton, *Contemporary debates in cognitive science*, 22–36.

Prinz, J. (2006b). Putting the brakes on enactive perception. http://www.theassc.org/files/assc/2627.pdf

Prinz, J. (2013). Siegel's get rich quick scheme. *Philosophical Studies: An International Journal for Philosophy in the Analytic Tradition, 163*(3), 827–835.

Prinz, J. J. (2012). *The conscious brain: How attention engenders experience*. Philosophy of mind series. Oxford: Oxford University Press.

Prinz, W., & Hommel, B. (Eds.). (2002). *Common mechanisms in perception and action:* Attention and performance. Oxford: Oxford University Press.

Putnam, H. (1975). *Mind, language and reality*. Philosophical Papers (Vol. 2). Cambridge, UK: Cambridge University Press.

Pylyshyn, Z. W. (1999). Vision and cognition: How do they connect? *Behavioral and Brain Sciences, 22*(3), 401–414.

Pylyshyn, Z. W. (2003). *Seeing and visualizing: It's not what you think*. Cambridge, Mass.: MIT Press.

Raichle, M. E. (2010). Two views of brain function. *Trends in Cognitive Sciences, 14*(4), 180–190. doi:10.1016/j.tics.2010.01.008.

Raichle, M. E., MacLeod, A. M., Snyder, A. Z., Powers, W. J., Gusnard, D. A., & Shulman, G. L. (2001). A default mode of brain function. *Proceedings of the National Acad-*

References

emy of Sciences of the United States of America, 98(2), 676–682. doi:10.1073/pnas.98.2.676.

Raichle, M. E., & Snyder, A. Z. (2007). A default mode of brain function: A brief history of an evolving idea. NeuroImage, 37(4), 1083–1090, discussion 1097–1099. doi:10.1016/j.neuroimage.2007.02.041.

Ramachandran, V. S., & Blakeslee, S. (1998). Phantoms in the brain: Probing the mysteries of the human mind. New York: William Morrow.

Rao, R. P., & Ballard, D. H. (1999). Predictive coding in the visual cortex: A functional interpretation of some extra-classical receptive-field effects. Nature Neuroscience, 2(1), 79–87. doi:10.1038/4580.

Raymond, J. E., Shapiro, K. L., & Arnell, K. M. (1992). Temporary suppression of visual processing in an RSVP task: An attentional blink? Journal of Experimental Psychology: Human Perception and Performance, 18(3), 849–860.

Rensink, R., O'Regan, J. K., & Clark, J. (1997). To see or not to see: The need for attention to detect changes in scenes. Psychological Science, 8(5), 368–373.

Richter, R. (2009). Der Skeptizismus in der Philosophie. BiblioBazaar. (1904).

Riffert, F. G., & Weber, M. (Eds.). (2002). Searching for new contrasts. Vienna: Peter Lang.

Riggs, L. A. (1952). The effects of counteracting the normal movements of the eye. Journal of the Optical Society of America, XLII, 872–873.

Rinzel, J., & Ermentrout, G. B. (1989). Analysis of neural excitability and oscillations. Methods in neuronal modeling 2: 251–292.

Rizzolatti, G., Luppino, G., & Matelli, M. (1998). The organization of the cortical motor system: New concepts. Electroencephalography and Clinical Neurophysiology, 106(4), 283–296. doi:10.1016/S0013-4694(98)00022-4.

Rizzolatti, G., Riggio, L., Dascola, I., & Umiltá, C. (1987). Reorienting attention across the horizontal and vertical meridians: Evidence in favor of a premotor theory of attention. Neuropsychologia, 25(1), 31–40. doi:10.1016/0028-3932(87)90041-8.

Robson, D. (2014). Neuroscience: The man who saw time stand still. http://www.bbc.com/future/story/20140624-the-man-who-saw-time-freeze/.

K. Rockland, J. Kass & A. Peters (Eds.). (1997). Cerebral cortex: Extrastriate cortex in primates. New York: Springer Science+Business Media.

Rockland, K. S., and van Hoesen, G W. (1994). Direct temporal-occipital feedback connections to striate cortex (V1) in the macaque monkey. Cerebral Cortex, 4(3), 300–313.

Rodemeyer, L. M. (2006). *Intersubjective temporality: It's about time.* Phaenomenologica 176. Dordrecht: Springer.

Rodriguez, E., George, N., Lachaux, J. P., Martinerie, J., Renault, B., & Varela, F. J. (1999). Perception's shadow: Long-distance synchronization of human brain activity. *Nature, 397*(6718), 430–433. doi:10.1038/17120.

Rorty, R. (1980). *Philosophy and the mirror of nature.* (2nd print., with corrections.) Princeton, N.J.: Princeton University Press.

Ross, D. and Ladyman, J. (2010). The alleged coupling-constitution fallacy and the mature sciences. In Menary, *The extended mind*, 155–166.

Rossetti, Y., Ota, H., Blangero, A., Vighetto, A., and Pisella, L. (2010). Why does the perception-action functional dichotomy not match the ventral-dorsal streams anatomical segregation: Optic ataxia and the function of the dorsal stream. In Gangopadhyay, Madary, and Spicer, *Perception, action, and consciousness*, 163–182.

Rossetti, Y., Pisella, L., & Vighetto, A. (2003). Optic ataxia revisited: Visually guided action versus immediate visuomotor control. *Experimental Brain Research, 153*(2), 171–179. doi:10.1007/s00221-003-1590-6.

Rossetti, Y., Revol, P., McIntosh, R., Pisella, L., Rode, G., Danckert, J., et al. (2005). Visually guided reaching: Bilateral posterior parietal lesions cause a switch from fast visuomotor to slow cognitive control. *Neuropsychologia, 43*(2), 162–177. doi:10.1016/j.neuropsychologia.2004.11.004.

Rowlands, M. (2010). *The new science of the mind: From extended mind to embodied phenomenology.* Cambridge, Mass.: MIT Press.

Rozzi, S., Calzavara, R., Belmalih, A., Borra, E., Gregoriou, G. G., Matelli, M., & Luppino, G. (2006). Cortical connections of the inferior parietal cortical convexity of the macaque monkey. *Cerebral Cortex, 16*(10), 1389–1417. doi:10.1093/cercor/bhj076.

Russell, B. (1978). *The problems of philosophy.* Oxford: Oxford University Press. (1912).

Sandin, R. H., Enlund, G., Samuelsson, P., & Lennmarken, C. (2000). Awareness during anaesthesia: A prospective case study. *Lancet, 355*(9205), 707–711. doi:10.1016/S0140-6736(99)11010-9.

Schellenberg, S. (2007). Action and self-location in perception. *Mind, 115,* 603–632.

Schellenberg, S. (2008). The situation-dependency of perception. *Journal of Philosophy, 105*(2), 55–84. doi:10.2307/20620076.

Schellenberg, S. (2011). Perceptual content defended. *Noûs, 45*(4), 714–750.

References

Schenk, T., Mai, N., Ditterich, J., & Zihl, J. (2000). Can a motion-blind patient reach for moving objects? *European Journal of Neuroscience, 12*(9), 3351–3360.

Schiller, P. H., Finlay, B. L. & Volman, S. F. (1976). Quantitative studies of single-cell properties in monkey striate cortex: V. Multivariate statistical analyses and models. *Journal of Neurophysiology, 39*(6), 1362–1374.

Schiller, P. H., & Logothetis, N. K. (1990). The color-opponent and broad-band channels of the primate visual system. *Trends in Neurosciences, 13*(10), 392–398. doi:10.1016/0166-2236(90)90117-S.

Schlick, M. (1938). Form and content: An introduction to philosophical thinking. In *Gesammelte Aufsätze 1926–1936*. Vienna: Gerold. 1938. Reprinted in Mulder & van de Velde-Schlick (Eds.) *Philosophical papers*. Dordrecht: Reidel, 1979.

Schmicking, D., & Gallagher, S. (2010). *Handbook of phenomenology and cognitive science*. Dordrecht: Springer.

Schroeder, T. (2006). Propositional attitudes. *Philosophy Compass, 1*, 65–73.

Schwender, D., Kunze-Kronawitter, H., Dietrich, P., Klasing, S., Forst, H., & Madler, C. (1998). Conscious awareness during general anaesthesia: Patients' perceptions, emotions, cognition and reactions. *British Journal of Anaesthesia, 80*(2), 133–139.

Schwitzgebel, E. (2006). Do things look flat? *Philosophy and Phenomenological Research, 72*(3), 589–599.

Schwitzgebel, E. (2008). The unreliability of naive introspection. *Philosophical Review, 117*(2), 245–273. doi:10.1215/00318108-2007-037.

Searle, J. R. (1983). *Intentionality: An essay in the philosophy of mind*. New York: Cambridge University Press.

Sellars, W., Rorty, R., & Brandom, R. (1997). *Empiricism and the philosophy of mind*. Cambridge, Mass.: Harvard University Press.

Sergent, C., & Dehaene, S. (2004). Is consciousness a gradual phenomenon? Evidence for an all-or-none bifurcation during the attentional blink. *Psychological Science, 15*(11), 720–728. doi:10.1111/j.0956-7976.2004.00748.x.

Seth, A. K. (2014). A predictive processing theory of sensorimotor contingencies: Explaining the puzzle of perceptual presence and its absence in synesthesia. *Cognitive Neuroscience, 5*(2), 97–118. doi:10.1080/17588928.2013.877880.

Shannon, B. (2003). Hallucinations. *Journal of Consciousness Studies, 10*, 3–31.

Shapiro, L. (Ed.). (2014). *The Routledge handbook of embodied cognition*. London: Routledge.

Sharp, D., & Leech, R. (2012). What role does the default mode network play in cognition? http://journal.frontiersin.org/researchtopic/575.

Sherman, S. M., & Guillery, R. W. (2002). The role of the thalamus in the flow of information to the cortex. *Philosophical Transactions of the Royal Society of London. Series B, Biological Sciences, 357*(1428), 1695–1708. doi:10.1098/rstb.2002.1161.

Shields, C. (2015). Aristotle's psychology. The Stanford Encyclopedia of Philosophy. http://plato.stanford.edu/archives/spr2015/entries/aristotle-psychology/.

Shipp, S., de Jong, B. M., Zihl, J., Frackowiak, R. S., & Zeki, S. (1994). The brain activity related to residual motion vision in a patient with bilateral lesions of V5. *Brain, 117*(Pt 5), 1023–1038.

Shulman, R. G. (2013). *Brain imaging: What it can (and cannot) tell us about consciousness.* Oxford: Oxford University Press.

Siegel, S. (2006). Which properties are represented in perception? In Gendler & Hawthorne, *Perceptual Experience*, 481–503.

Siegel, S. (2010a). *The contents of visual experience.* Oxford: Oxford University Press.

Siegel, S. (2010b). Do experiences have contents? In Nanay, *Perceiving the world,* 333–368.

Siegel, S. (2015). The contents of perception. The Stanford Encyclopedia of Philosophy. http://plato.stanford.edu/archives/spr2015/entries/perception-contents/.

Siewert, C. (2005). Attention and sensorimotor intentionality. In Smith & Thomasson, *Phenomenology and philosophy of mind,* 270–294.

Siewert, C. P. (1998). *The significance of consciousness.* Princeton, N.J: Princeton University Press.

Silveira, L. C. L., & Perry, V. H. (1991). The topography of magnocellular projecting ganglion cells (M-ganglion cells) in the primate retina. *Neuroscience, 40*(1), 217–237. doi:10.1016/0306-4522(91)90186-R.

Simons, P. (1995). Meaning and language. In Smith & Smith, *The Cambridge companion to Husserl,* 106–137.

Simons, D., & Chabris, C. (1999). Gorillas in our midst: Sustained inattentional blindness for dynamic events. *Perception, 28*(9), 1059–1074.

Smith, A. D. (2000). Space and sight. *Mind, 109*(435), 481–518.

Smith, A. D. (2002). *The problem of perception.* Cambridge, Mass.: Harvard University Press.

Smith, A. D. (2003). *Routledge philosophy guidebook to Husserl and the Cartesian meditations.* Routledge philosophy guidebooks. London: Routledge.

Smith, B., & Smith, D. W. (Eds.). (1995). *The Cambridge companion to Husserl.* Cambridge, UK: Cambridge University Press.

Smith, D. W., & McIntyre, R. (1982). *Husserl and intentionality: A study of mind, meaning, and language.* Synthese library (Vol. 154). Dordrecht, Holland: D. Reidel.

Smith, D. W., & Thomasson, A. L. (Eds.). (2005). *Phenomenology and philosophy of mind.* Oxford: Oxford University Press.

Smith, J. (2010). Seeing other people. *Philosophy and Phenomenological Research, 81*(3), 731–748. doi:10.1111/j.1933-1592.2010.00392.x.

Snow, C. P. (1959). *The two cultures.* Cambridge, U.K.: Cambridge University Press.

Snowden, R. J., Treue, S., & Andersen, R. A. (1992). The response of neurons in areas V1 and MT of the alert rhesus monkey to moving random dot patterns. *Experimental Brain Research, 88*(2), 389–400.

Snowdon, P. (1980–1981). Perception, vision and causation. *Proceedings of the Aristotelian Society, 81,* 175–192.

Softky, W. R., & Koch, C. (1993). The highly irregular firing of cortical cells is inconsistent with temporal integration of random EPSPs. *Journal of Neuroscience, 13*(1), 334–350.

Sokolowski, R. (1970). *The formation of Husserl's concept of constitution.* Phaenomenologica (Vol. 18). The Hague: Martinus Nijhoff.

Soteriou, M. (2014). The disjunctive theory of perception. The Stanford Encyclopedia of Philosophy. http://plato.stanford.edu/archives/sum2014/entries/perception-disjunctive/.

Sperling, G. (1960). The information available in brief visual presentations. *Psychological Monographs, 74*(11), 1–29.

Spiegelberg, H. (1978). *The phenomenological movement: A historical introduction.* 2nd ed. Phaenomenologica (Vol. 1). The Hague: Martinus Nijhoff.

Spivey, M. (2007). *The continuity of mind.* Oxford psychology series 44. New York: Oxford University Press.

Sporns, O. (2011). *Networks of the brain.* Cambridge, Mass.: MIT Press.

Sporns, O., & Zwi, J. D. (2004). The small world of the cerebral cortex. *Neuroinformatics, 2*(2), 145–162. doi:10.1385/NI:2:2:145.

Sprague, J. M. (1966). Interaction of cortex and superior colliculus in mediation of visually guided behavior in the cat. *Science, 153*(3743), 1544–1547.

Spratling, M. W. (2008). Predictive coding as a model of biased competition in visual attention. *Vision Research, 48*(12), 1391–1408. doi:10.1016/j.visres.2008.03.009.

Spratling, M. W. (2012). Predictive coding accounts for V1 response properties recorded using reverse correlation. *Biological Cybernetics*, *106*(1), 37–49. doi:10.1007/s00422-012-0477-7.

Spratling, M. W. (2012). Predictive coding as a model of the V1 saliency map hypothesis. *Neural networks 26*, 7–28. doi:10.1016/j.neunet.2011.10.002.

Stainton, J. (Ed.). (2006). *Contemporary debates in cognitive science*. Wiley-Blackwell.

Stazicker, J. (2011). Attention, visual consciousness and indeterminacy. *Mind & Language*, *26*(2), 156–184.

Steinbock, A. (2003). Generativity and the scope of generative phenomenology. In Welton, *The new Husserl*, 289–326.

Steinbock, A. J. (1995). *Home and beyond: Generative phenomenology after Husserl*. Northwestern University studies in phenomenology and existential philosophy. Evanston, Ill.: Northwestern University Press.

Stepanyants, A., Martinez, L. M., Ferecskó, A. S., & Kisvárday, Z. F. (2009). The fractions of short- and long-range connections in the visual cortex. *Proceedings of the National Academy of Sciences of the United States of America*, *106*(9), 3555–3560. doi:10.1073/pnas.0810390106.

Stephen, D. G., & Dixon, J. A. (2011). Strong anticipation: Multifractal cascade dynamics modulate scaling in synchronization behaviors. *Chaos, Solitons, and Fractals*, *44*(1–3), 160–168. doi:10.1016/j.chaos.2011.01.005.

Stephen, D. G., Stepp, N., Dixon, J. A., & Turvey, M. T. (2008). Strong anticipation: Sensitivity to long-range correlations in synchronization behavior. *Physica A*, *387*(21), 5271–5278. doi:10.1016/j.physa.2008.05.015.

Stokes, D. (2014). Cognitive penetration and the perception of art. *Dialectica*, *68*(1), 1–34. doi:10.1111/1746-8361.12049.

Stokes, D., Matthen, M., & Biggs, S. (Eds.). (2014). *Perception and its modalities*. Oxford: Oxford University Press.

Strogatz, S. H. (2000). *Nonlinear dynamics and chaos: With applications to physics, biology, chemistry, and engineering*. Studies in nonlinearity. Cambridge, MA: Westview Press.

Swoyer, C. (1995). Leibnizian expression. *Journal of the History of Philosophy*, *33*(1), 65–99.

Tappen, M., Freeman, W., & Adelson, E. (2005). Recovering intrinsic images from a single image. *IEEE Transactions on Pattern Analysis and Machine Intelligence*, *27*(9), 1459–1472.

Taylor, J. (1962). *The behavioral basis of perception*. New Haven, CT: Yale University Press.

Thelen, E., & Smith, L. B. (1996). *A dynamic systems approach to the development of cognition and action*. Cambridge, Mass.: MIT Press.

Thompson, B. (2006). Color constancy and Russellian representationalism. *Australasian Journal of Philosophy*, 84(1), 75–94.

Thompson, E., Noë, A., & Pessoa, L. (1999). Perceptual completion: A case study in phenomenology and cognitive science. In Petitot, *Naturalizing phenomenology*, 161–195.

Thompson, E. (2007). *Mind in life: Biology, phenomenology, and the sciences of mind*. Cambridge, Mass.: Belknap.

Todd, J., Norman, J. F., & Mingolla, E. (2004). Lightness constancy in the presence of specular highlights. *Psychological Science*, 15(1), 33–39.

Tolhurst, D. J., Movshon, J. A., & Dean, A. F. (1983). The statistical reliability of signals in single neurons in cat and monkey visual cortex. *Vision Research*, 23(8), 775–785.

Tomasello, M. (1999). *The cultural origins of human cognition*. Cambridge, Mass.: Harvard University Press.

Tomasello, M., Carpenter, M., Call, J., Behne, T., & Moll, H. (2005). Understanding and sharing intentions: The origins of cultural cognition. *Behavioral and Brain Sciences*, 28(5), 675–691, discussion 691–735. doi:10.1017/S0140525X05000129.

Travis, C. (2004). The silence of the senses. *Mind*, 113(449), 57–94.

Travis, C. (2013). *Perception: Essays after Frege*. Oxford: Oxford University Press.

Treisman, A. M., & Gelade, G. (1980). A feature-integration theory of attention. *Cognitive Psychology*, 12(1), 97–136.

Treisman, A. (1986). Features and objects in visual processing. *Scientific American*, 255(5), 114–125. doi:10.1038/scientificamerican1186-114B.

Treue, S., & Maunsell, J. H. (1996). Attentional modulation of visual motion processing in cortical areas MT and MST. *Nature*, 382(6591), 539–541. doi:10.1038/382539a0.

Trevarthen, C. (1979). Communication and cooperation in early infancy. In Bullowa, *Before speech*, 321–348.

Tribus, M. (1961). *Thermodynamics and thermostatistics: An introduction to energy, information and states of matter, with engineering applications*. New York: D. van Nostrand.

Troncoso, X. G., Macknik, S. L., Otero-Millan, J., & Martinez-Conde, S. (2008). Microsaccades drive illusory motion in the Enigma illusion. *Proceedings of the National Academy of Sciences of the United States of America, 105*(41), 16033–16038. doi:10.1073/pnas.0709389105.

Troost, J. M., & de Weert, C. M. (1991). Naming versus matching in color constancy. *Perception & Psychophysics, 50*(6), 591–602.

Tsuda, I. (2001). Toward an interpretation of dynamic neural activity in terms of chaotic dynamical systems. *Behavioral and Brain Sciences, 24*(05), 793. doi:10.1017/S0140525X01000097.

Tye, M. (2000). *Consciousness, color, and content.* Cambridge, Mass.: MIT Press.

Tye, M. (2009). A new look at the speckled hen. *Analysis, 69*, 258–263.

Tye, M. (2010). Up close with the speckled hen. *Analysis, 70*(2), 283–286.

Von Uexküll, J. (1934). A stroll through the worlds of animals and men. In Lashley, *Instinctive behavior*, 5–80.

Van Essen, D. C. (2004). Organization of visual areas in macaque and human cerebral cortex. In Chalupa and Werner, *The visual neurosciences*, 507–521.

Van Gelder, T. (1995). What might cognition be, if not computation? *Journal of Philosophy, 92*(7), 345. doi:10.2307/2941061.

Van Gulick, R. (2007). What if phenomenal consciousness admits of degrees? *Behavioral and Brain Sciences, 30*(5–6), 528–529.

Van Orden, G., Pennington, B., & Stone, G. (2001). What do double dissociations prove? *Cognitive Science, 25*, 111–172.

Varela, F. (1999). The specious present: A neurophenomenology of time consciousness. In Petitot, *Naturalizing phenomenology*, 266–314.

Varela, F., Lachaux, J. P., Rodriguez, E., & Martinerie, J. (2001). The brainweb: Phase synchronization and large-scale integration. *Nature Reviews Neuroscience, 2*(4), 229–239. doi:10.1038/35067550.

Varela, F. J. (1996). Neurophenomenology: A methodological remedy for the hard problem. *Journal of Consciousness Studies, 3*(4), 330–349.

Varela, F. J., Thompson, E., & Rosch, E. (1991). *The embodied mind: Cognitive science and human experience.* Cambridge, Mass.: MIT Press.

Vision, G. (1997). *Problems of vision: Rethinking the causal theory of perception.* New York: Oxford University Press.

References

Vogels, R., Spileers, W., & Orban, G. A. (1989). The response variability of striate cortical neurons in the behaving monkey. *Experimental Brain Research, 77*(2), 432–436. doi:10.1007/BF00275002.

Stein, W. (1989). *On the problem of empathy.* Washington, D.C.: ICS. (1913).

Wallhagen, M. (2007). Consciousness and action: Does cognitive science support (mild) epiphenomenalism? *British Journal for the Philosophy of Science, 58*(3), 539–561. doi:10.1093/bjps/axm023.

Ward, D., & Stapleton, M. (2012). Es are good: Cognition as enacted, embodied, embedded, affective and extended. In Paglieri, *Consciousness in interaction,* 89–104.

Watts, D. J., & Strogatz, S. H. (1998). Collective dynamics of "small-world" networks. *Nature, 393*(6684), 440–442. doi:10.1038/30918.

Weddell, R. A. (2004). Subcortical modulation of spatial attention including evidence that the Sprague effect extends to man. *Brain and Cognition, 55*(3), 497–506. doi:10.1016/j.bandc.2004.02.075.

Welton, D. (Ed.). (2003). *The new Husserl: A critical reader.* Studies in Continental thought. Bloomington, IN: Indiana University Press.

Wexler, M. (2005). Anticipating the three-dimensional consequences of eye movements. *Proceedings of the National Academy of Sciences of the United States of America, 102*(4), 1246–1251. doi:10.1073/pnas.0409241102.

Williamson, T. (1982). Intuitionism disproved? *Analysis, 42*(4), 203–207. doi:10.1093/analys/42.4.203.

Williamson, T. (2000). *Knowledge and its limits.* Oxford: Oxford Univ. Press.

Wissig, S. C., & Kohn, A. (2012). The influence of surround suppression on adaptation effects in primary visual cortex. *Journal of Neurophysiology, 107*(12), 3370–3384. doi:10.1152/jn.00739.2011.

Yarbus, A. (1967). *Eye movements and vision.* New York: Plenum Press.

Yoshimi, J. (2009). Husserl's theory of belief and the Heideggerean critique. *Husserl Studies, 25*(2), 121–140. doi:10.1007/s10743-008-9046-2.

Yoshimi, J. (2011). Phenomenology and connectionism. *Frontiers in Psychology, 2*, 288. doi:10.3389/fpsyg.2011.00288.

Yoshimi, J. (2012a). Supervenience, dynamical systems theory, and non-reductive physicalism. *British Journal for the Philosophy of Science, 63*(2), 373–398. doi:10.1093/bjps/axr019.

Yoshimi, J. (2012b). Active internalism and open dynamical systems. *Philosophical Psychology, 25*(1), 1–24. doi:10.1080/09515089.2011.569919.

Yoshimi, J. (2015). The metaphysical neutrality of Husserlian phenomenology. *Husserl Studies*, *31*(1), 1–15. doi:10.1007/s10743-014-9163-z.

Yoshimi, J. (2016). *Husserlian phenomenology: A unifying interpretation*. Springer briefs in philosophy. Switzerland: Springer.

Zahavi, D. (2001). Beyond empathy: Phenomenological approaches to intersubjectivity. *Journal of Consciousness Studies*, *8*(5–7), 151–167.

Zahavi, D. (2005). *Subjectivity and selfhood: Investigating the first-person perspective*. Cambridge, Mass.: MIT Press.

Zahavi, D. (2008). Internalism, externalism, and transcendental idealism. *Synthese*, *160*(3), 355–374. doi:10.1007/s11229-006-9084-2.

Zeimbekis, J., & Raftopoulos, A. (Eds.). (2015). *The cognitive penetrability of perception*. Oxford: Oxford University Press.

Zeki, S. (1991). Cerebral akinetopsia (visual motion blindness). A review. *Brain*, *114*(Pt 2), 811–824.

Zemel, R. (1993). A minimum description length framework for unsupervised learning. Ph.D. thesis, University of Toronto, Toronto.

Zemel, R., & Hinton, G. (1995). Learning population codes by minimizing description length. *Neural Computation*, *7*(3), 549–564.

Zihl, J., von Cramon, D., & Mai, N. (1983). Selective disturbance of movement vision after bilateral brain damage. *Brain*, *106*(2), 313–340. doi:10.1093/brain/106.2.313.

Index

2.5D sketch, 12–16, 18, 24–25, 27
4E cognitive science, 163

Abstraction problem, the, 72–73
Accuracy conditions, 59, 67–68, 163
Action, as related to perception, 5–12, 22, 99–103, 173–176
Active vision, 16, 38, 49, 63, 92, 109
AF, thesis, 3–4, 8–10, 26, 29, 33, 38, 107, 145, 155–156, 158, 164–165, 167, 171, 176–177, 180–183
AF content, 59–65, 69–79, 83–86, 167
Affordances, 93–94, 175
Afterimage, 40, 47, 146
Aglioti, S., 142
Akinetopsia (motion blindness), 145–149
Akins, K., 199n
Amodal completion, 105, 108
Anatomical localization, 17–18, 22, 24
Anderson, M., 19, 23
Anticipation. *See* Visual anticipation
Anti-representational, 160, 163
Area MT/V5, Brain, 20, 137, 146, 149
Arieli, A., 122
Aristotle, 5, 119, 134
Atmanspacher, H., 161
Attention, visual, 51, 53, 56, 72, 78, 82, 85–86, 97–98, 111–117, 125, 128, 172, 186, 188
  covert, 113–114
  endogenous, 112
  exogenous, 53, 112
  as increased determinacy of visual anticipation, 111–116
  overt, 113–114
  as precision optimization, 111–114
  pre-motor theory of, 111–114
Attentional blink, 81, 115–117
Austin, J. L., 71
Autism, 172
Autopoiesis, 120
Axon, 121
Ayer, A. J., 28

Backpropagation, 126
Ballard, D., 100–101, 127–128
Bar, M., 124, 150–151
Barber, M., 85
Bayesian inference, 94–95, 112–114, 126, 159. *See also* Predictive processing
Bayne, T., 79, 185–188
Bechtel, W., 25, 120, 123, 160
Belief, 5, 11, 41, 51–53, 56–57, 65–66, 68, 83–87, 97, 104–107, 171, 174
  perceptual justification of, 83–87 (*see also* Intelligible interface problem; Davidson/McDowell worry)
Bickhard, M., 120
Block, N., 17, 22, 26, 44, 79–83, 133–134, 139

Brain
  active, 23, 25, 119–123
  default mode network, 122, 128
  endogenous dynamics of, 25, 119–121
  noise in, 121–122
Brain damage, 21, 24, 106, 108, 138, 146, 151, 175
Brain imaging, 23, 120, 122. *See also* fMRI
Breitmeyer, B., 150
Brewer, B., 43, 71–74
Breyer, T., 188
Broca's area, 19
Bruner, J., 93, 104, 110
Bullier, J., 151
Burge, T., 66–67, 199n, 201n

Callaway, E., 136–137
Carruthers, P., 133–134, 157
Cartesian theater, 22
Casler, K., 171, 176
Category theory, 86, 188
Causation/constitution fallacy, 22
Chalmers, D., 157, 184
Chameleon effect, 173
Change blindness, 16, 34, 96–98
Clark, A., 25, 95, 120, 126, 129, 159, 164, 199n
Classical sandwich, 5–6, 9–10, 13, 174
Cognitive penetration, 110
Color, 29–31, 36, 50, 61, 73–74, 82–83, 101–102, 104, 110, 115–116, 136–137, 140, 146, 166, 178–179, 192n. *See also* Perspectival color
Constitution, 22, 180–182
Context-sensitive
  neural response, 18–19, 122–123, 129
  visual content, 73, 93, 123, 129
Crane, T., 67–69
Crick, F., 124
Cross-cultural comparison, 172

Cycle
  of action and perception, 5–6, 9–10, 12, 57, 135, 173–176
  of anticipation and fulfillment, 22, 40, 135, 148–149

Dacey, D., 138
Davidson, D., 66–68, 75, 83–85, 114, 162. *See also* Davidson/McDowell worry
Davidson/McDowell worry, 68, 75, 83–85
Dendrite, 121
Dennett, D., 11–12, 18, 21–22, 34, 36, 37, 45, 56, 67, 78, 81, 101–102, 135–136, 140, 156, 175–176
  card trick, 36–37, 48, 78, 140
Descartes, R., 5, 189
D.F., patient, 139–141
Dialogue, between humanities and sciences, 157
Disjunctivism, 70–74. *See also* Perceptual content, denial of
Distinction between perception and cognition, 56–57
Dorsal stream, 16, 131–134, 136–142, 144, 149–151
Doyon, M., 85, 194
Dretske, F., 44, 59, 70, 75–76, 97
Dual visual systems, 128, 131–138
Dummett, M., 67
Dynamic point-light display, 172

Early vision module (EVM), 103–108, 110
Embodied stance, the, 175–176
Enactivism, 8–9, 84, 120, 133
Endogenous neural activity, 25, 119–121
Erregen. See Visual anticipation, stirred up
Error signal, 113, 115, 126–127. *See also* Predictive processing

Index

Evolution, 97, 150
  cultural, 173
Explanatory gap, xii, 188
Extrastriate cortex, 12, 14–16, 19–20, 23–24

Factual content, 43–47, 60–61, 71, 159
Factual properties, 28–29, 32, 38–39, 42, 44–47, 50–53, 60–61, 65, 70, 73, 76, 92, 160–162, 178
  incomplete perception of, 38, 61, 71, 76, 97, 170
Familiarity with a visual scene, 40, 50, 53, 55–56, 95, 111, 163, 167–168, 170
Feedback, 10, 105–110, 149
  neural, 20, 95, 119, 123–129, 150–151
Flowerpot with miniature city, 51–53
fMRI, 19, 21–23, 192n
Fodor, J., 5, 9–10, 21, 59, 70, 94, 97, 103–110, 124, 162, 174
Fovea, 16, 36, 78, 132, 135–136, 138, 141, 149, 151
Freeman, W., 125, 128
Front-loaded phenomenology, 108, 151
Functionalism, 13

Gärdenfors, P., 161
Generative model, 113–114, 125–126. *See also* Predictive processing
Gibson, J., 6, 93–94, 175
Goggles, inverting, 99, 102–103
Goodale, M. *See* Milner, D. and Goodale, M.
Good experience (of the doll), the, 42–43, 74
Grasping, visually guided, 139, 142–144
Gregory, R., 92–93
Gurwitsch, A., 38

Hallucination, visual, 47, 62–64, 70, 110
Hallucinogen-persisting perception disorder, 145–146

Hallucinogens, 145–146
Hard problem of consciousness, xi, 185, 188
Held, R. and Hein, A., 16, 102
Higher-order thought theories of consciousness, 117, 133
Hinton, G., 125–126
Hobbes, T., 189
Hohwy, J., 25, 94, 110, 113, 126, 158–159
Hopp, W., 29, 85, 179
Horizon, visual, 132–136, 139–142, 144, 150, 151, 181
Hubel, D., 122–124, 127, 136–137
Hume, D., 28, 32, 188
Hurley, S., 5–6, 9–11, 21–22, 34, 99, 174–175
Hutto, D., 84, 162. *See also* Radical enactive cognition
Husserl, E., 4–5, 28, 34, 36, 47–50, 60–61, 71, 78–79, 85–86, 108, 132–136, 139–140, 151, 165–167, 177–189
Hypothesis testing, perception as, 92

Illusion, 34, 51, 62–64, 70, 105
  hollow mask, 93
  Müller-Lyer, 62–64, 105, 107
  rotating snakes (peripheral drift), 143–144
  Titchener circles (Ebbinghaus), 143
Imitation, 175
Inattentional blindness, 16, 96–98, 115, 117
Indeterminate, visual experience is, 16, 25, 27, 34, 36–40, 45, 48, 55, 62, 69, 71, 77–83, 85, 95–98, 103, 109, 117, 132, 134–136, 140, 178, 186
Information encapsulation, 21
Initial view of a scene, 39–40, 49–52
Intelligible interface problem, 84

Intention (as used by Husserl)
 fulfilled, 177–179
 partial, 178–180
 total, 178–179
Intentional stance, 11–12, 175–176
Introspection, 77–79, 156, 185–187. *See also* Skepticism

Jackendoff, R. 4, 12, 14, 157. *See also* Jackendoff/Prinz argument
Jackendoff/Prinz argument, 12, 14–16, 21, 25–26, 35, 129, 157, 185
James, W., 33, 36, 185
Judgment, perceptual, 63, 74, 77, 83–86, 144, 186

Kanizsa, G., 105–107
Kant, I., 32, 181
Kelly, S. D., 29, 73
Kitaoka, A., 143–144
Koch, C., 121, 124
Kohler, I., 16, 102
Kyoto School, 185

Laasik, K., 135, 177, 180–184, 197n
Land, M., 100, 102, 109
Lê, S., 140–141
Lee, T. S., 128
Leibniz, G., 28, 32
Lewis, D., 13, 54–55
Livingstone, M., 136–137
Llinás, R., 121–122, 125, 128
L.M., patient, 146, 149
Local field potential, 120, 122
LSD, 145

Magnocellular pathway, 136–138, 141, 150–151
Main Argument, 3–4, 15, 17, 26, 38, 57–58, 87, 91, 93, 111, 117, 146, 155–160, 164, 168
Marr, D., 12–18, 21, 25, 27, 35, 104, 124, 163

Matthen, M., 69
McDowell, J., 66, 68, 70, 75, 83–85
Merleau-Ponty, M., 5, 38, 165, 168–169, 182
Metabolism, 23, 119
Metzinger, T., 33, 44, 47, 199n
Microsaccade, 49, 99
Migraine, 146
Millikan, R., 199n
Milner, D. *See* Milner, D. and Goodale, M.
Milner, D. and Goodale, M., 16, 131–134, 137–142, 149
Mismatch negativity, 160
Modal, 4
Modularity
 horizontal, 9–11
 vertical, 9, 191n
Montague, M., 67–68
Moore, G. E., 28, 43, 113
Movement, self-generated. *See* Self-generated movement
Mulligan, K., 194n, 200n
Mumford, D., 128
Myin, E., 84, 162. *See also* Radical enactive cognition
Myth of full detail, 96–98, 103

Nagel, T., 156–158
Nanay, B., 26, 38, 75–76
Nassi, J., 136–137
Natural language, 62, 65–69, 71, 78, 83–84, 97, 174
Neural connectivity, 10, 18, 20–21, 123, 125
Neural network, 24, 125–127
Neuron, 18–21, 94, 120–122, 124, 127–128
 context sensitivity of response, 18–20, 121–122, 127
 pacemaker, 121
 resonator, 121
 single-cell recording of, 19, 121–122

Neurophenomenology, 177, 184
"New Look" psychology, 93, 104
Noë, A., 6–7, 9, 20, 27–30, 34, 38, 44, 64, 94, 99, 103, 133, 147, 180
Norms of tool use, 167–168, 171, 175–176
No-strong-loops hypothesis, 140

Object-relational view of perceptual content, 68–69
Odd experience (of the doll), the, 42–43, 74
Öğmen, H., 150
Optic ataxia, 139, 141–142
Ordinary language philosophy, 71
O'Regan, K., 7, 16, 96, 99, 101, 103, 180

Palinopsia, 145–148
  causes of, 146
Paralysis, 7–8, 49, 64–65
Parvocellular pathway, 137–138, 150–151
Peacocke, C., 28, 59, 66, 69, 70
Perceptual content
  conceptual, 84–85, 140
  denial of, 43, 70–75
  nonconceptual, 73, 85
  variation in, 54–56
Perceptual propositionalism, 65–68, 70–71, 74, 86
Peripheral vision, 16, 25, 36–38, 48, 69, 78, 83, 96–99, 109, 112, 132, 135, 141–144
Personal/subpersonal distinction, 56–57, 95, 114, 158–160
Perspectival, visual experience is, 15, 25, 27–32, 38–39, 71, 74, 170, 178, 183, 186
  properties, 28–29, 32, 44
Perspectival color, 30–31, 178–179
Perspectival connectedness (PC), 41–45, 47–48, 50–51, 53, 59, 74, 184

Perspectival connectedness' (PC'), 42–43, 45–49, 51, 53–55, 57–61, 64–65
Petersen, B., 138
Petitot, J., 86, 184, 188, 191n
Phenomenalism, 61, 73
Phenomenal overflow, 79–81
Philips, I., 81
Physicalism, 182
Pisella, L., 132, 141–142
Posner, M., 114–115
Possible worlds, 4, 66
Predictive processing, 25, 94–95, 112–115, 123, 125–129, 150, 158–160, 163. *See also* Bayesian inference; Generative model
Premise
  descriptive, 3–4, 15, 26, 38, 57, 87, 155–156
  empirical, 3–4, 14–15, 17, 26, 91–92, 94–95, 99, 103, 112, 117, 120, 131, 151, 155–156
Prinz, J., 4, 12–25, 117, 124, 170, 185. *See also* Jackendoff/Prinz argument
Projected size, perception of, 29–30
Propositional attitudes, 5, 10, 66–68, 160, 174
Psychophysics, perceptual, 29, 118
Pylyshyn, Z., 94, 96–97, 104–106, 124

Radical enactive cognition, 84. *See also* Hutto, D.; Myin, E.
Raichle, M., 120, 122
Rao, R., 127–128
Rationality, 10, 84, 107, 174–175
  rational behavior, 11–12, 176
  rational justification, 85–87
Receptive field, 94, 122, 124, 127, 137
  extra-classical response of, 94, 127
Representation, mental, 9, 12–15, 24, 29, 32, 37–39, 45–46, 59, 61, 70–71, 76, 92, 103, 109, 118, 155, 160–164, 175, 191n

Retention of fulfilled anticipation, 60, 147–148
Retina, 7, 13, 16, 25, 36, 54, 78, 99, 103, 132, 135–138
  distribution of photoreceptors in, 16, 36, 138
Retinocortical dynamics model (RECOD), 150
Rorty, R., 83–86
Rossetti, Y., 132, 141–143
Russell, B., 28, 43, 66

Saccade, 6–7, 35, 38, 54–56, 60–63, 76–78, 82, 96–102, 109, 112–116, 163, 172
  task-dependence of, 100–101
S.B., patient, 140–141
Scenario content, 65, 69
Schellenberg, S., 6, 28–29, 34–35, 44, 54–55
Schlick, M., 157
Schwitzgebel, E. 26, 29, 77–79, 156
Searle, J., 44, 66
Selection problem, the, 72–73
Selective rearing, 16, 99, 102–103
Self-generated movement, 8, 39, 42–45, 47, 51–52, 56, 60, 64, 73, 92
  visual disturbance of, 198n
Seizure, 124, 146
Sense-datum theory, 43
Sensorimotor
  approach, 7, 94, 101, 133, 179–180
  contingencies, 46, 69, 94, 103, 159
  loop, 6, 174–175
Seth, A., 94–95, 159
Shared intentionality, 173
Siegel, S., 40–43, 47, 50–53, 56, 59, 66, 74, 183, 194n
Siewert, C., 27, 29, 33, 46, 48, 56
Skepticism, 26, 77–79, 181, 186–187
Small-world connectivity, 18, 21
Smith, J., 166–168

Social cognition, 11, 164–165, 170–172, 174–176
  disorders of, 172
Social content of vision, 3, 11, 165–176
Social signals in visual perception, 173
Sokolowski, R., 182
Soul, 119
Spatial fringe, 132, 134–135, 140
Spatiotemporal processing differences of two visual streams, 16, 132–134, 138, 144, 149
Specific anticipation (SA), 42–46, 53, 59
Speckled hen, the problem of the, 26, 75–76
Specular highlights, 31–32, 55
Sperling, G., 79–82
Spivey, M., 19, 161–162
Spratling, M., 115, 128
Stazicker, J., 81–82
Stein, E., 166–168, 188
Steinbock, A., 135, 187
Stream of consciousness, 33–34, 147
Surprisal, 159–160
Surprise, 37–38, 45–47, 97, 108, 159–160, 191n
Symbolic dynamics, 155, 160–162, 188

Temporal fringe, 123, 135
Temporal, visual experience is, 25, 27, 29, 32–36, 38–39, 69–70, 76, 79, 147
Thalamus, 136, 150
Thermodynamic equilibrium, 120
Thought experiment
  the child artificially raised, 169
  the doll on the shelf, 42–43, 74, 183
  perfect hallucination, 64
  the predicting contest, 12
Thompson, E., 20, 38, 120, 162, 166, 184
Tomasello, M., 173
Top-down neuroscience, 17, 20–23
Trailing, visual, 145–147
Transcendental idealism, 181–182

Index

Trans-saccadic perception, 25, 35–36, 147
Travis, C., 43, 70–72, 74
Triesman, A., 115
Truth conditions, 67
Tye, M., 30, 44, 75–76, 129, 191n
Two visual systems. *See* Dual visual systems

Unconscious inference, perception as, 92
Unconscious priming, 117
Unification problem, the, 34–35. *See also* Schellenberg, S.

Varela, F., 120, 128, 162, 177, 184–185, 188
Ventral stream, 16, 131–134, 136–142, 149–151
Veridicality, 62–64, 70–73
Verschmolzen. See Visual anticipation, fusion of
Vestibular, 40
Virtual reality, 33, 108, 193n
Vision
  central, 36, 48, 98, 103, 138, 141–144
  peripheral, 16, 25, 36–38, 48, 69, 78, 81–83, 96–99, 109, 112, 132, 135, 141–144 (*see also* Indeterminate, visual experience is)
  snapshot conception of, 27, 33, 146–148, 198n
"Vision for action," 16, 131–134, 139, 142, 149
"Vision for perception," 16, 131, 133–134, 139, 142, 149
Visual agnosia, 106, 108, 138, 144–145
Visual anticipation
  five features of, 48
  fusion of, 61–62, 178
  implicit, 31, 39–40, 45, 48–49, 63–64, 76, 96, 108, 112, 135, 176, 178
  indeterminate, 40, 48, 55, 82–83, 95, 98, 136 (*see also* Indeterminate, visual experience is)
  scope of, 51–52
  stirred up, 47–48, 50, 52, 60, 61, 63, 72–73, 78, 82, 98, 107, 11–112, 116, 118, 173, 178–179
von Helmholtz, H., 92

Wallhagen, M., 140
Wiesel, T., 122–124, 127
Wittgenstein, L., 38

Yarbis, A., 56, 99–100
Yoshimi, J., 85–86, 161, 178, 182, 188